HILARY W. K. CHEN

*Multistage Transistor Circuits*

ITU Library
Date:

INTERNATIONAL TECHNOLOGICAL UNIVERSITY
This Book is Donated by:
PROF. WAI-KAI CHEN

Date:

## Semiconductor Electronics Education Committee Books

**Vol. 1** **Introduction to Semiconductor Physics**
R. B. Adler, A. C. Smith, and R. L. Longini

**Vol. 2** **Physical Electronics and Circuit Models of Transistors**
P. E. Gray, D. DeWitt, A. R. Boothroyd, and J. F. Gibbons

**Vol. 3** **Elementary Circuit Properties of Transistors**
C. L. Searle, A. R. Boothroyd, E. J. Angelo, Jr., P. E. Gray, and D. O. Pederson

**Vol. 4** **Characteristics and Limitations of Transistors**
R. D. Thornton, D. DeWitt, E. R. Chenette, and P. E. Gray

**Vol. 5** **Multistage Transistor Circuits**
R. D. Thornton, C. L. Searle, D. O. Pederson, R. B. Adler, and E. J. Angelo, Jr.

**Vol. 6** **Digital Transistor Circuits**
J. N. Harris and P. E. Gray

**Vol. 7** **Handbook of Basic Transistor Circuits and Measurements**
R. D. Thornton, J. G. Linvill, E. R. Chenette, H. L. Ablin, J. N. Harris, A. R. Boothroyd, and J. Willis

# Multistage Transistor Circuits

*Semiconductor Electronics Education Committee, Volume 5*

Richard D. Thornton
Massachusetts Institute of Technology

Campbell L. Searle
Massachusetts Institute of Technology

Donald O. Pederson
University of California, Berkeley

R. B. Adler
Massachusetts Institute of Technology

E. J. Angelo, Jr.
Polytechnic Institute of Brooklyn

*John Wiley & Sons, Inc., New York · London · Sydney*

*Copyright © 1965 by Educational Services, Inc.*

All Rights Reserved. This book or any part thereof must not be reproduced in any form without the written permission of the publisher.

*Library of Congress Catalog Card Number: 65-19471*

*Printed in the United States of America*

# Foreword

The importance of transistors and other semiconductor devices is now well established. The subsequent development of microminiaturized electronic circuits has blurred the dividing line between the "device" and the "circuit," and thus has made it increasingly important for us to understand deeply the relationship between the internal physics and structure of a device, and its potentialities for circuit performance. Furthermore, the small size and efficient operation of semiconductor devices make possible for the first time a much closer integration between the theoretical and laboratory aspects of the educational process.

To prepare new educational material which would reflect these developments, there was formed in the Fall of 1960 a group known as the Semiconductor Electronics Education Committee (SEEC). This committee is comprised of university and industrial members, brought together by several of the faculty of the Electrical Engineering Department at the Massachusetts Institute of Technology, with Professor C. L. Searle acting as Chairman and Professor R. B. Adler acting as Technical Director. The committee undertook the production of a multipurpose course in semiconductor electronics, designed primarily for use in universities at the third or fourth year undergraduate level.

The success of the high-school physics course developed by the Physical Science Study Committee (PSSC) led the SEEC to believe that the same kind of combination used there—text, laboratory experiments, and films, in a complementary format—would be the most practical way of providing uniformly high-quality instruction over the wide range of material involved. It was hoped that this arrangement would lead to broad applicability of the course in the academic world, and also in some professional training activities of industry and government. This book is one in the SEEC series, all volumes of which are listed here:*

Vol. 1 (ISP) *Introduction to Semiconductor Physics*, R. B. Adler, A. C. Smith, and R. L. Longini

Vol. 2 (PEM) *Physical Electronics and Circuit Models of Transistors*, P. E. Gray, D. DeWitt, A. R. Boothroyd, and J. F. Gibbons

Vol. 3 (ECP) *Elementary Circuit Properties of Transistors*, C. L. Searle, A. R. Boothroyd, E. J. Angelo, Jr., P. E. Gray, and D. O. Pederson

Vol. 4 (CLT) *Characteristics and Limitations of Transistors*, R. D. Thornton, D. DeWitt, E. R. Chenette, and P. E. Gray

Vol. 5 (MTC) *Multistage Transistor Circuits*, R. D. Thornton, C. L. Searle, D. O. Pederson, R. B. Adler, and E. J. Angelo, Jr.

Vol. 6 (DTC) *Digital Transistor Circuits*, J. N. Harris and P. E. Gray

Vol. 7 (TCM) *Handbook of Basic Transistor Circuits and Measurements*, R. D. Thornton, J. G. Linvill, E. R. Chenette, H. L. Ablin, J. N. Harris, A. R. Boothroyd, and J. Willis

These books have all gone through at least one "preliminary edition," many through two or more. The preliminary editions were used in teaching trials at some of the participating colleges and industrial training activities, and the results have been used as a basis for revision.

* Minor changes in title or authorship may take place in some of the volumes, which are still in preparation at the time of this writing.

It is almost impossible to enumerate all those people who have contributed some of their effort to the SEEC. Certain ones, however, have either been active with the Committee steadily since its inception, or have made very major contributions since then. These may be thought of as "charter members," deserving special mention.

*From Universities*
    California, Berkeley: D. O. Pederson
    Imperial College, London: A. R. Boothroyd[Δ]
    Iowa State: H. L. Ablin[*]
    M.I.T.: R. B. Adler, P. E. Gray, A. L. McWhorter, C. L. Searle, A. C. Smith, R. D. Thornton, J. R. Zacharias, H. J. Zimmermann (Research Laboratory of Electronics), J. N. Harris (Lincoln Laboratory)
    Minnesota: E. R. Chenette
    New Mexico: W. W. Grannemann
    Polytechnic Institute of Brooklyn: E. J. Angelo, Jr.
    Stanford: J. F. Gibbons, J. G. Linvill
    U.C.L.A.: J. Willis

*From Industries*
    Bell Telephone Laboratories: J. M. Early, A. N. Holden, V. R. Saari
    Fairchild Semiconductor: V. R. Grinich
    IBM: D. DeWitt
    RCA: J. Hilibrand, E. O. Johnson, J. I. Pankove
    Transitron: B. Dale,[†] H. G. Rudenberg[‡]
    Westinghouse Research Laboratories: A. I. Bennett, H. C. Lin, R. L. Longini[§]

General management of the SEEC operations is in the hands of Educational Services, Inc. (abbreviated ESI), Watertown, Mass.,

[Δ] Now at Queen's University, Belfast.
[*] Now at the University of Nebraska, Department of Electrical Engineering.
[†] Now at Sylvania Corp.
[‡] Now at A. D. Little, Inc.
[§] Now at Carnegie Institute of Technology, Department of Electrical Engineering.

a nonprofit corporation that grew out of the PSSC activities and is presently engaged in a number of educational projects at various levels. In addition to providing general management, ESI has supplied all the facilities necessary for preparing the SEEC films. These are 16-mm sound films, 30 to 40 minutes in length, designed to supplement the subject matter and laboratory experiments presented in the various text books. The film titles are:

"Gap Energy and Recombination Light in Germanium"— J. I. Pankove and R. B. Adler

"Minority Carriers in Semiconductors"—J. R. Haynes and W. Shockley

"Transistor Structure and Technology"—J. M. Early and R. D. Thornton.

Pending arrangements for commercial distribution, these films are available (purchase or rental) directly from Educational Services, Inc., 47 Galen Street, Watertown, Mass.

The committee has also endeavored to develop laboratory materials for use with the books and films. This material is referred to in the books and further information about it can be obtained from ESI.

The preparation of the entire SEEC program, including all the books, was supported at first under a general grant made to the Massachusetts Institute of Technology by the Ford Foundation, for the purpose of aiding in the improvement of engineering education, and subsequently by specific grants made to ESI by the National Science Foundation. This support is gratefully acknowledged.

Campbell L. Searle
*Chairman*, SEEC
Richard B. Adler
*Technical Director*, SEEC

# *Preface*

This book, the fifth in the Semiconductor Electronics Education Committee series, is concerned primarily with the problems in multistage transistor amplifiers operating in the linear region. The topic is obviously broad, so broad in fact that complete coverage within the scope of one book is clearly impossible. The authors believed, however, that certain material was essential to the understanding of the subject. For convenience, this material has been gathered together in the first four chapters to form Part I of this book. These four chapters contain two important topics: methods of analysis of multistage amplifiers, with emphasis on methods of handling the interaction between transistor stages; and approximate methods for analyzing transistor feedback amplifiers.

For convenience of teaching, each of the remaining four chapters, which comprise Part II, have been written so as to depend only on the material in Part I. Thus the teacher can feel free to select any chapter or group of chapters from Part II to illustrate some of the more detailed aspects of multistage amplifiers.

We have assumed in writing this book that the reader has a detailed working knowledge of calculations of gain and bandwidth by using the hybrid-$\pi$ model; also a knowledge of how the transistor parameters vary with voltage, current, and temperature. These facts are essential for a true understanding of most of the design discussions, and are examined in detail in the first three volumes of the SEEC series.

The notation for transistor terminal variables used in the SEEC textbooks conforms to the IEEE standard. That is, subscripts indicate the terminal at which the current flows or the terminal pair at which a voltage appears, and the nature of the signal is indicated as follows:

    dc or operating point variables—upper-case symbols with upper-case subscripts

    total instantaneous variables—lower-case symbols with upper-case subscripts

    incremental instantaneous variables—lower-case symbols with lower-case subscripts

    complex amplitudes of incremental components—upper-case symbols with lower-case subscripts

The authors are indebted to all the members of the SEEC who have, over the past three years, helped to shape the content of this book. We are also indebted to Willis North, John Kassakian, and James Maskasky, students at M.I.T., who helped in preparing the illustrations, doing the experimental work, and reading proof. The cooperation of the town of Wayland in allowing us to use the facilities of the Wayland High School during the summers of 1961, 1962, and 1963 is gratefully acknowledge.

R. D. THORNTON<br>
C. L. SEARLE<br>
D. O. PEDERSON<br>
R. B. ADLER<br>
E. J. ANGELO, JR.

# Contents

**Part One  General Background, 1**

**1  Gain and Bandwidth Calculations:
A First Approximation, 3**

1.0  Introduction, 3
1.1  Mid-Frequency Gain Calculation, 6
1.2  Bandwidth Calculations, 8
1.3  Finding Approximate Pole Locations, 18
1.4  Use of the $C_t$ Approximation to Find $\omega_h$ and the Lowest Pole, 22
1.5  Extension of the $C_t$ Approximation to Find $A_v(j\omega)$, 26
1.6  Behavior at High Frequency, 30

**2  Amplifier Calculations Using a π Model, 33**

2.0  Introduction, 33
2.1  Transistor Model Simplification by Absorbing $r_x$, 35
2.2  Illustrative Example: Three-Stage Amplifier, 41
2.3  Performance Calculations Based on Transfer Function, 45
2.4  Simplified Model Including $R_e{'}$, 52
2.5  Summary, 56

## 3 Feedback Amplifier Concepts, 62

3.0 Introduction, 62
3.1 Elementary Properties of Feedback Amplifiers, 65
3.2 More Detailed Formulation of Feedback Viewpoint, 74
3.3 Examples of Elementary Feedback Analysis, 91

## 4 Stability and Frequency Response of Feedback Amplifiers, 104

4.0 Introduction, 104
4.1 The Stability Problem, 104
4.2 Root Locus Calculations, 108
4.3 Feedback Amplifier Design Based on $j$-Axis Response, 125

### Part Two Selected Topics, 145

## 5 Broadband Amplifiers, 147

5.0 Introduction, 147
5.1 Typical Interior Stage of a Cascaded Amplifier, 148
5.2 Resistive Broadbanding, 150
5.3 Broadbanding by Addition of Emitter Resistance, 158
5.4 The Shunt-Peaked Amplifier, 163

## 6 Direct-Coupled Amplifiers, 170

6.0 Introduction and Basic Considerations, 170
6.1 Typical Circuit Configurations, 172
6.2 Effects of Parameter Changes, 185
6.3 Symmetric Circuits, 201

## 7 Tuned Multistage Amplifiers, 224

7.0 Introduction, 224
7.1 Cascaded Neutralized Stages, 225
7.2 Single-Stage Building Block for Multistage Amplifier without Neutralization, 228
7.3 Multistage Amplifiers with Interaction, 233
7.4 Illustrative Design Examples, 236

## 8 Interrelations Between Frequency Domain, Time Domain, and Circuit Parameters, 244

8.0 Introduction, 244
8.1 Direct Calculation of Frequency and Time Response, 245
8.2 Approximations for $H(j\omega)$ and $h(t)$, 251
8.3 Low-Frequency Considerations, 274

*Index, 283*

*Part One*

*General Background*

# 1

## Gain and Bandwidth Calculations: A First Approximation

**1.0 INTRODUCTION**

Significant new design possibilities appear when we progress from the single-stage circuits discussed in Elementary Circuit Properties of Transistors* to multistage circuits. For example, we can achieve very large gains when transistor stages are cascaded (that is, the output of one stage is connected to the input of the second stage), because the system gain will usually be of the order of the product of the gains of the individual stages. However, accompanying these new design possibilities are new design problems. For example, the problem of interaction between various stages in multistage amplifiers becomes of major importance in the calculation of high-frequency behavior.

In this chapter, we bring out the problems of the interaction among coupled transistor stages, by discussing the properties of a small-signal, low-pass amplifier. We wish to focus the discussion primarily on the calculation of the bandwidth of the amplifier; thus many other important considerations, such as biasing, tem-

---

* *Elementary Circuit Properties of Transistors*, by C. L. Searle, A. R. Boothroyd, E. J. Angelo Jr., P. E. Gray, and D. O. Pederson, hereafter referred to as ECP.

perature stability, and coupling and bypass capacitor calculations (and, hence, low-frequency behavior), will not be treated.

Our method of attack will be first to calculate the mid-frequency gain of the amplifier shown in Fig. 1.1, then to make an increasingly accurate series of calculations of the amplifier frequency response at high frequencies. We start by estimating, on the basis of the intrinsic hybrid-$\pi$ model, the frequency where the magnitude of the voltage gain is down to 0.707 of its mid-frequency value. (This frequency is referred to as the 0.707 frequency, or the 3-db frequency, or the half-power frequency, or just the 0.707 point.) We then calculate the approximate location of the lowest and highest poles of the transfer function, and finally make a first approximation to the magnitude and phase of $A_v(j\omega)$. In Chapter 2, more accurate estimates are made of the pole locations and frequency response by converting from the hybrid-$\pi$ model to a $\pi$ model, and computing the fourth-order denominator polynomial for this circuit. The results of all these calculations are intercompared in the course of Chapter 2.

We assume that the three transistors used in the amplifier are silicon epitaxial *npn* units with the following parameter values, measured at the operating point $V_{CE} = 5$ volts, $I_C = 5$ milliamperes:

$f_T = 400$ mc  $\qquad r_\pi = 400$ ohms

$g_m = 0.2$ mho  $\qquad r_x = 50$ ohms

$C_\pi = 78.5$ pf  $\qquad r_\mu = 1$ megohm

$C_\mu = 2.5$ pf  $\qquad r_o = 12$ k

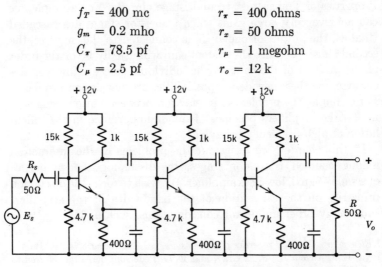

Fig. 1.1. Schematic diagram of three-stage transistor amplifier.

If we choose to represent each of the three transistors by the intrinsic hybrid-$\pi$ model given in Fig. 1.2, we could obtain the complete small-signal circuit model for the three-stage amplifier. However, it is probably wiser to simplify the model somewhat before proceeding further. Following the method outlined in Chapter 6 of ECP, we will develop two models: one will allow us to calculate the mid-frequency gain by inspection, and the other will allow us to calculate the high-frequency behavior.

Most $RC$ amplifiers are designed to have a constant gain characteristic over some frequency range. Below this region, the gain will fall off because of the effects of coupling and bypass capacitors. Above this region, the gain will fall because of the effects of $C_\pi$, $C_\mu$, stray shunt capacitance, etc. If the amplifier has a substantial frequency range with constant gain—more than one or two decades, say—then it is easy to identify the capacitors which contribute to the low-frequency behavior and those which contribute to high-frequency behavior, and thus construct separate low-frequency and high-frequency circuit models. Said in another way, under these conditions, we can split the transfer function into two factors:

$$A_v(s) = \left[\frac{N_l(s)}{(s-s_1)(s-s_2)\ldots}\right]\left[\frac{N_h(s)}{(s-s_\alpha)(s-s_\beta)\ldots}\right] \quad (1.1)$$

in which the first factor, containing the low-frequency poles and zeros, is nearly independent of $C_\pi$, $C_\mu$, $C_{bc}$, etc., and the second factor, containing the high-frequency poles and zeros, is nearly independent of the coupling and bypass capacitor values.

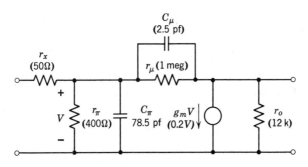

Fig. 1.2. Hybrid-$\pi$ model.

## 6 Gain and Bandwidth Calculations: A First Approximation

On this basis, we draw in Fig. 1.3 the mid-frequency and high-frequency small-signal circuit models for the three-stage amplifier. Throughout this chapter we completely ignore the low-frequency performance of the amplifier, because this part of multistage analysis can be handled adequately by the methods given in ECP, Chapter 6. In the figure, the 1-k load resistor and the two biasing resistors have been combined in each interstage network into one equivalent resistor, $R_i$, where

$$R_i = R_c \parallel R_{b1} \parallel R_{b2} \tag{1.2}$$

(Two parallel vertical lines indicate "in parallel with.") Note also that the input-stage biasing resistors have been omitted in each circuit model because they are so large compared to the 50-ohm value of $R_s$.

In this design, the value of the total resistive load on each stage is less than 300 ohms, so it is easy to see that the gain per stage must be less than $0.2 \times 300 = 60$. This gain is sufficiently low to allow us to neglect both $r_\mu$ and $r_o$. Because we can neglect $r_\mu$, the circuit in Fig. 1.3a is an adequate mid-frequency small-signal representation of the amplifier, that is, from several hundred kc down to a frequency where the coupling capacitors and bypass capacitors can no longer be considered as short circuits.

The high-frequency model of Fig. 1.3b is valid up to approximately the transverse cutoff frequency $\omega_b$:

$$\omega_b = \frac{g_x + g_\pi}{C_\pi + C_\mu} = \frac{22.5}{81} \times 10^9 \tag{1.3}$$

$$= 0.28 \times 10^9 \text{ rads/sec.}$$

$$= 45 \text{ mc}$$

### 1.1 MID-FREQUENCY GAIN CALCULATION

The calculation of the mid-frequency gain from Fig. 1.3a is straightforward. The over-all voltage gain expression is:

$$A_{v0} = \frac{V_o}{E_s}\bigg|_{\text{mid-frequency}}$$

$$= \left(\frac{-r_\pi g_m R_{L1}}{r_\pi + r_x + R_s}\right)\left(\frac{-r_\pi g_m R_{L2}}{r_\pi + r_x}\right)\left(\frac{-r_\pi g_m R_{L3}}{r_\pi + r_x}\right) \tag{1.4}$$

### Sec. 1.1 Mid-Frequency Gain Calculation 7

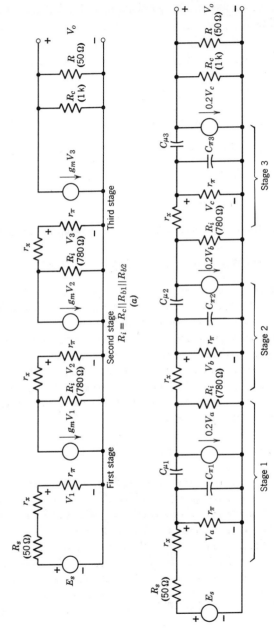

**Fig. 1.3.** Circuit models for three-stage amplifier. (*a*) Small-signal mid-frequency circuit model. (*b*) Small-signal high-frequency circuit model (valid up to 45 mc).

where $R_L$ is in each case the total resistive load on the stage in question. That is,

$$R_{L1} = R_i \| (r_x + r_\pi) \tag{1.5}$$

$$R_{L2} = R_{L1} \tag{1.6}$$

$$R_{L3} = R_c \| R \tag{1.7}$$

Using the numerical values from Figs. 1.1 and 1.2, we find

$$R_i = 1 \text{ k} \| 4.7 \text{ k} \| 15 \text{ k} = 780 \text{ }\Omega \tag{1.8}$$

$$R_{L1} = R_{L2} = 780 \| 450 = 285 \text{ }\Omega \tag{1.9}$$

$$R_{L3} = 1 \text{ k} \| 50 \text{ }\Omega \cong 50 \text{ }\Omega \tag{1.10}$$

Therefore, the over-all mid-frequency voltage gain of the amplifier is:

$$A_{v0} = -(45.5)(51)(8.9) = -21{,}000 \tag{1.11}$$

Incidentally, it is clear from Eq. 1.11 just how the gain of the amplifier would change if additional identical transistor stages were added between the second and third stage of the original amplifier. The gain would increase by a factor of about 50 for each additional stage added. The number of such stages that can be cascaded successfully is determined primarily by the noise of the input stage and the saturation characteristics of the output stage. Certainly, cascading must stop before the noise of the input stage alone saturates the output stage.

## 1.2 BANDWIDTH CALCULATIONS

Now let us determine the pass band of the amplifier or, more accurately, the high frequency at which the magnitude of the voltage gain of the amplifier falls to $0.707 |A_{v0}|$. The small-signal circuit model appropriate for these calculations is given in Fig. 1.3b.

The difficulty in bandwidth calculations for multistage circuits can be traced directly to the feedback capacitor $C_\mu$. This capacitor provides coupling back from the output to the input of each transistor, thus preventing us from treating each transistor as a unilateral device. The intercoupling caused by these capacitors must be accounted for in most designs, particularly when the best present-day transistors are used.

The exact analysis of the circuit in Fig. 1.3b can be handled in several ways. For example, we could solve the six equations corresponding to the six nodes of the network. (The seventh node can be eliminated by combining $R_s$ and $r_x$ at the input.) However, in this chapter we wish to develop much simpler methods of analysis which will give us a first approximation to the 0.707 point of the amplifier.

**1.2.1** *Finding the 0.707 Point from the First Two Denominator Terms of the Transfer Function*

Our first approximate analysis method is based on the fact that the high-frequency 0.707 point often can be calculated with considerable accuracy from only the first two or three terms in the denominator of the transfer function. To show this, we first find the general form of the voltage-gain expression $V_o/E_s$ for Fig. 1.3b. This function will have six poles corresponding to the six independent energy-storage elements in the circuit. Inspection of the circuit reveals no poles or zeros at zero frequency, and a third-order zero at infinite frequency resulting from the three capacitors $C_\pi$. Thus the transfer function will have the form:

$$\frac{V_o}{E_s} = \frac{K'(s - s_7)(s - s_8)(s - s_9)}{(s - s_1)(s - s_2)(s - s_3)(s - s_4)(s - s_5)(s - s_6)} \quad (1.12)$$

As we shall see, all the poles of this function lie on the negative real axis. Three of the poles occur at $|s|$ less than $\omega_T$, and three occur at $|s|$ greater than $\omega_T$. The three zeros arise from the forward coupling through $C_\mu$ in each transistor cancelling the signal from the current source $g_m V$, just as in the single-stage case. Thus these will be right half-plane zeros, located at $s = g_m/C_\mu$ for each stage, a value of $|s|$ much larger than $\omega_T$ for that stage. In this example, all these zeros are at the same frequency: $s_7 = s_8 = s_9 = g_m/C_\mu$. They will obviously be of little consequence in determining the 0.707 frequency of the amplifier, because the 0.707 frequency will surely be much lower in frequency than $g_m/C_\mu$. Accordingly, we neglect $s$ in comparison with $s_7$, $s_8$, and $s_9$ in the numerator of Eq. 1.12, and thus obtain an approximate gain expression in which these zeros have been moved out to infinity:

$$\frac{V_o}{E_s} = \frac{K}{(s-s_1)(s-s_2)(s-s_3)(s-s_4)(s-s_5)(s-s_6)}$$

$$= \frac{K}{a_0 + a_1 s + a_2 s^2 + a_3 s^3 + \ldots s^6} \quad (1.13)$$

The 0.707 frequency, as we said, is that frequency $s = j\omega_h$ where $|V_o/E_s| = 0.707 |K|/a_0$. Substituting $j\omega_h$ for $s$ in Eq. 1.13 and taking magnitudes, we obtain

$$\left|\frac{V_o}{E_s}\right| = \frac{|K|}{\sqrt{2}\, a_0} = \frac{|K|}{|a_0 + ja_1\omega_h - a_2\omega_h^2 - ja_3\omega_h^3 \ldots|} \quad (1.14)$$

Inverting and finding the squared magnitude, we obtain

$$2a_0^2 = (a_0 - a_2\omega_h^2 + \ldots)^2 + \omega_h^2(a_1 - a_3\omega_h^2 \ldots)^2 \quad (1.15a)$$

$$= a_0^2 - 2a_0 a_2 \omega_h^2 + a_1^2 \omega_h^2 - 2a_1 a_3 \omega_h^4 + \ldots \quad (1.15b)$$

If we keep terms out to $\omega_h^2$ in Eq. 1.15b, we may solve for $\omega_h$ to obtain for the upper 0.707 point:

$$\omega_h \cong \frac{a_0}{\sqrt{a_1^2 - 2a_0 a_2}} \quad (1.16)$$

This kind of approximation turns out to be best when all of the poles lie close to or on the negative real axis, as they do in this type of circuit. (This approximation is discussed in more detail in Chapter 8.)

Often for real-axis poles we can make an even cruder approximation to $\omega_h$ by neglecting the $a_2$ term as well:

$$\omega_h \cong \frac{a_0}{a_1} \quad (1.17)$$

That is, *the upper 0.707 frequency can be evaluated approximately from only the constant term and the coefficient of s in the denominator polynomial.* Equation 1.17 is particularly useful because often the ratio of $a_0$ to $a_1$ can be found by inspection of the network, whereas except for quadratics and cubics, coefficient $a_2$ usually requires substantially more calculation. We discuss this point in more detail in the following subsection.

For the case where all of the poles lie on the negative real axis, it is a relatively simple matter to assess the accuracy of Eq. 1.17. We first write the amplifier transfer function for $s = j\omega$ in the form:

$$A_v(j\omega) = \frac{A_{v0}}{\left(1 - \frac{j\omega}{s_1}\right)\left(1 - \frac{j\omega}{s_2}\right)\left(1 - \frac{j\omega}{s_3}\right)\cdots\left(1 - \frac{j\omega}{s_n}\right)} \quad (1.18)$$

where $s_1, s_2 \ldots s_n$ are *real* constants because we are assuming all real poles. Multiplying out, we find:

$$a_0 = 1 \quad (1.19)$$

$$a_1 = -\left(\frac{1}{s_1} + \frac{1}{s_2} + \frac{1}{s_3} \cdots + \frac{1}{s_n}\right) \quad (1.20)$$

If the poles are widely separated on the real axis, that is, if $|s_1| \ll |s_2| \ll |s_3| \ldots \ll |s_n|$, then it is clear from Eq. 1.18 that $s_1$ will control the upper 0.707 point. Specifically, we would expect $\omega_h \cong |s_1|$. Applying Eq. 1.17 to the present case, we find from Eqs. 1.19 and 1.20:

$$\omega_h \cong -s_1 \quad (1.21)$$

Thus for the case of widely separated poles, Eq. 1.17 gives the 0.707 point without significant error.

The reader can convince himself by working some examples that the error introduced by Eq. 1.17 is greatest when the poles all occur at the same frequency, i.e., when $s_1 = s_2 = s_3 \ldots = s_n$, or:

$$A(j\omega) = \frac{A_0}{\left(1 - \frac{j\omega}{s_1}\right)^n} \quad (1.22)$$

For this case, Eqs. 1.17, 1.19, and 1.20 predict an upper 0.707 frequency of:

$$\omega_h \cong \frac{|s_1|}{n} \quad (1.23)$$

## 12 Gain and Bandwidth Calculations: A First Approximation

The actual frequency $\omega_h'$ at which the gain is reduced by $\sqrt{2}$, is given by

$$\sqrt{1 + (\omega_h'/s_1)^2} = (\sqrt{2})^{1/n} \qquad (1.24)$$

or

$$\omega_h' = |s_1| \sqrt{2^{1/n} - 1} \qquad (1.25)$$

The frequencies $\omega_h'$ and $\omega_h$ and the percentage error for various values of $n$ are shown in Table 1.1.

**TABLE 1.1**

| $n$ | $\omega_h = \dfrac{\|s_1\|}{n}$ | $\omega_h'' = \dfrac{\|s_1\|}{\sqrt{n}}$ | $\omega_h' = \|s_1\| \sqrt{2^{1/n} - 1}$ | $100\left(\dfrac{\omega_h - \omega_h'}{\omega_h'}\right)$ |
|---|---|---|---|---|
| 1 | $\|s_1\|$ | $\|s_1\|$ | $\|s_1\|$ | 0 |
| 2 | $0.500\ \|s_1\|$ | $0.707\ \|s_1\|$ | $0.644\ \|s_1\|$ | $-22.4\%$ |
| 3 | $0.333\ \|s_1\|$ | $0.576\ \|s_1\|$ | $0.510\ \|s_1\|$ | $-34.7\%$ |
| 4 | $0.250\ \|s_1\|$ | $0.500\ \|s_1\|$ | $0.435\ \|s_1\|$ | $-42.6\%$ |

In multistage amplifiers, it is unusual to find multiple-order poles, because the natural result of the interaction between stages is to split the poles apart. The error introduced by our approximation for $\omega_h$ decreases very rapidly when this splitting occurs. For example, with three poles the error is, at worst, about $-6\%$ if the second and third poles are at least a decade above the first. Note that for identical real-axis poles Eq. 1.16 gives $\omega_h'' = |s_1|/\sqrt{n}$, which is closer to the true value of $\omega_h$ than Eq. 1.17. In this case the latter expression will always give a value for the 0.707 frequency which is *below* the true value.

Thus we have developed a method for calculating the approximate 0.707 point from the coefficients of the *unfactored* denominator polynomial of the transfer function. In the following section we relate these coefficients to various time constants in the network, thereby permitting evaluation of $a_1/a_0$ and $a_{n-1}/a_n$ by inspection of the network. By using these relations we can find approximate 0.707 points, and, as we shall see in Sec. 1.3, approximate pole locations as well, *without calculating the entire denominator polynomial*. More accurate approximate methods are discussed in Chapter 8, along with a discussion of time-domain response.

### ▶1.2.2 Finding $a_1/a_0$ and $a_{n-1}/a_n$ by Inspection of the Network*

Consider a *linear active* network which contains three capacitors and no other energy storage. We represent this network as shown in Fig. 1.4a, where each capacitor has been pulled out to form a terminal pair.

We subsequently have need for the values of the low-frequency short-circuit conductance and the open-circuit resistance seen at each terminal pair. To find these, we remove the capacitors, as shown in Fig. 1.4b, and represent the linear active network which remains by the node equations given in Eqs. 1.26. Any *independent* sources within the box are assumed to be set to zero.

$$I_1 = g_{11}V_1 + g_{12}V_2 + g_{13}V_3$$
$$I_2 = g_{21}V_1 + g_{22}V_2 + g_{23}V_3 \quad (1.26)$$
$$I_3 = g_{31}V_1 + g_{32}V_2 + g_{33}V_3$$

For convenience in the following discussion we will use the shorthand notation of determinants. In this case, the $g$ determinant is:

$$\Delta_g \equiv \begin{vmatrix} g_{11} & g_{12} & g_{13} \\ g_{21} & g_{22} & g_{23} \\ g_{31} & g_{32} & g_{33} \end{vmatrix} \quad (1.27)$$

By definition of the $y$-parameters, the conductance seen at terminal pair $j$ with all other terminal pairs *shorted* is $g_{jj}$.

We designate the resistance seen at terminal pair $j$ with all other terminal pairs *open* as $R_{jo}$. This open-circuit resistance can be evaluated by solving Eqs. 1.26 to obtain the voltage $V_j$ in terms of the current $I_j$ with all other currents set equal to zero. The result is, from Cramer's rule,

$$V_j = I_j \frac{G_{jj}}{\Delta_g} \quad (1.28)$$

where $G_{jj}$ denotes the *cofactor* of the element $g_{jj}$.† Thus $R_{jo} = V_j/I_j$ is

$$R_{jo} = \frac{G_{jj}}{\Delta_g} \quad (1.29)$$

---

* In an introductory course, sections marked with ▶ at the start and ◀ at the end, and set in smaller type, may be omitted without loss of continuity. The important results of the proof in this section are summarized at the beginning of Sec. 1.2.3.

† The cofactor $G_{jj}$ of an element on the principal diagonal is equal to the determinant which remains when the $j$th row and $j$th column of the $g$ determinant are deleted.

# 14 Gain and Bandwidth Calculations: A First Approximation

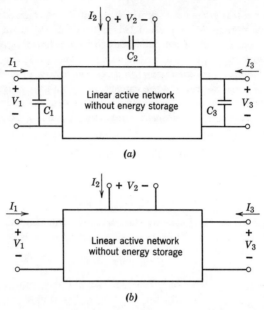

**Fig. 1.4.** Representation of a linear active network containing three capacitors. (a) Capacitors identified by terminal pair. (b) Network with capacitors removed.

When the capacitors are present at the terminal pairs, as shown in Fig. 1.4a, the complete network can be represented by a set of node equations with the following admittance determinant, which is obtained by adding the appropriate capacitive susceptances to the conductances on the principal diagonal of $\Delta_g$:

$$\Delta_y = \begin{vmatrix} g_{11} + sC_1 & g_{12} & g_{13} \\ g_{21} & g_{22} + sC_2 & g_{23} \\ g_{31} & g_{32} & g_{33} + sC_3 \end{vmatrix} \quad (1.30)$$

The *natural frequencies* of the complete network are those frequencies at which voltages can exist at the terminals when $I_1 = I_2 = I_3 = 0$. This condition can exist only if $\Delta_y = 0$. Inspection of Eq. 1.30 shows that $\Delta_y$ will contain, in general, all powers of $s$ from $s^0$ up to $s^3$. If there are $n$ capacitors and no capacitor loops, the highest power of $s$ in the characteristic equation is $n$. That is, the characteristic equation of the network must be of the form:

$$\Delta_y = 0 = a_0 + a_1 s + a_2 s^2 + a_3 s^3 \tag{1.31}$$

If all natural frequencies have negative real parts, the network is stable, and there then will be no missing coefficients in the middle of the equation. That is, in our example with three capacitors, $a_3$ can be zero if there is a capacitor loop, but if $a_3 \neq 0$ and $a_0 \neq 0$, then $a_1$ and $a_2$ must also be nonzero.

Some of the $a$'s can be identified relatively easily by inspection of $\Delta_y$. The coefficient $a_0$, the constant term, is simply $\Delta_y$ when $s = 0$, which is the same as when all the capacitors are removed. That is,

$$a_0 = \Delta_g \tag{1.32a}$$

The coefficient $a_1$ (that of the first power of $s$) must be the sum of three terms, each of which contains one capacitance multiplied by the cofactor of the corresponding element in the *conductance* determinant. On this basis, $a_1$ is:

$$a_1 = C_1 G_{11} + C_2 G_{22} + C_3 G_{33} \tag{1.32b}$$

The coefficient $a_3$ (or in general the coefficient of the $s^n$ term) results from multiplying the susceptive parts of the elements on the principal diagonal and is:

$$a_3 = C_1 C_2 C_3 \tag{1.32c}$$

Finally, the coefficient $a_2$ (in general the coefficient of the $s^{n-1}$ term) which also results from multiplying terms on the principal diagonal, contains three terms, each of which involves all capacitors except one:

$$a_2 = C_2 C_3 g_{11} + C_1 C_3 g_{22} + C_1 C_2 g_{33} \tag{1.32d}$$

Thus the coefficient $a_1/a_0$ is, from Eqs. 1.32a and 1.32b:

$$\frac{a_1}{a_0} = C_1 \frac{G_{11}}{\Delta_g} + C_2 \frac{G_{22}}{\Delta_g} + C_3 \frac{G_{33}}{\Delta_g} \tag{1.33}$$

However, $G_{jj}/\Delta_g$ is the open-circuit resistance seen by $C_j$, as shown by Eq. 1.29. Consequently,

$$\frac{a_1}{a_0} = R_{1o} C_1 + R_{2o} C_2 + R_{3o} C_3 \equiv \sum_j \tau_{jo} \tag{1.34}$$

where $\tau_{jo}$ denotes the time constant of the $j$th capacitor calculated with all other capacitors *open-circuited*. Open-circuiting all capacitors but one usually breaks a complicated network into small pieces, so the open-circuit time constants of a network can usually be found *by inspection*.

Equation 1.34 is the desired relationship for finding from Eq. 1.17 the high-frequency 0.707 point of the amplifier. We shall also derive at this point a second relation among the coefficients of the denominator polynomial, which is useful for finding the low-frequency 0.707 point of an amplifier, and, as we shall see in Sec. 1.3, the highest natural frequency of an amplifier. The desired relation, the ratio $a_2/a_3$, is, from Eqs. 1.32c and 1.32d,

$$\frac{a_2}{a_3} = \frac{g_{11}}{C_1} + \frac{g_{22}}{C_2} + \frac{g_{33}}{C_3} \quad (1.35)$$

$$\equiv \sum \frac{1}{\tau_{js}} \quad (1.36)$$

where $\tau_{js}$ denotes the time constant of the $j$th capacitor calculated with all other capacitors *shorted*. Shorting all capacitors but one usually collapses a complicated network sufficiently that again these time constants can be determined *by inspection*.

The results of this section can readily be generalized to networks with more than three capacitors. The general form of the characteristic equation (Eq. 1.31) for a network with $n$ capacitors is:

$$\Delta_y = 0 = a_0 + a_1 s + \ldots + a_{n-1} s^{n-1} + a_n s^n \quad (1.37)$$

The $a_1$ term now becomes:

$$a_1 = C_1 G_{11} + C_2 G_{22} + \ldots + C_n G_{nn} \quad (1.38)$$

and the ratio of the second to first terms becomes:

$$\frac{a_1}{a_0} = R_{1o}C_1 + R_{2o}C_2 + \ldots + R_{no}C_n \equiv \sum_j \tau_{jo} \quad (1.39)$$

Then $s^n$ coefficient becomes:

$$a_n = C_1 C_2 C_3 \ldots C_n \quad (1.40)$$

and the ratio of second-last to last coefficient becomes:

$$\frac{a_{n-1}}{a_n} = \frac{g_{11}}{C_1} + \frac{g_{22}}{C_2} + \ldots + \frac{g_{nn}}{C_n} \equiv \sum_j \frac{1}{\tau_{js}} \quad (1.41)$$

The coefficients of the remaining terms can also be found by somewhat similar methods, as discussed in Chapter 8. ◄

### 1.2.3 Example: Calculation of 0.707 Point

In the previous section we showed that for a network with $n$ capacitors, and no capacitor loops, the ratio of the coefficient of the $s$ term to the constant term in the denominator of the gain expression is:

$$\frac{a_1}{a_0} = R_{1o}C_1 + R_{2o}C_2 + \ldots + R_{no}C_n \equiv \sum_j \tau_{jo} \qquad (1.39)$$

where $R_{jo}$ denotes the resistance seen by the $j$th capacitor, and $\tau_{jo}$ denotes the time constant of the $j$th capacitor, both calculated with all other capacitors *open-circuited*. Also, we showed that the ratio of the coefficient of the $s^{n-1}$ term to the coefficient of the $s^n$ term is

$$\frac{a_{n-1}}{a_n} = \frac{g_{11}}{C_1} + \frac{g_{22}}{C_2} + \ldots + \frac{g_{nn}}{C_n} \equiv \sum_j \frac{1}{\tau_{js}} \qquad (1.41)$$

where $g_{jj}$ is the conductance facing $C_j$, and $\tau_{js}$ is the time constant of the $j$th capacitor, both calculated with all other capacitors *shorted*. Because both $\tau_{jo}$ and $\tau_{js}$ can often be found by inspection of the network, Eqs. 1.39 and 1.41 are very useful for first-order circuit calculations. For example, we see from Eqs. 1.39 and 1.17 that the approximate upper 0.707 frequency of $A_v$ is:

$$\omega_h \cong \frac{1}{\sum \tau_{jo}} \qquad (1.42)$$

Let us use Eq. 1.42 to calculate the bandwidth of the amplifier in Fig. 1.3b. Starting from the first stage, we calculate the open-circuit time constant involving $C_\pi$:

$$\tau_{1o} = C_{\pi 1}R_{1o} \qquad (1.43)$$

where $R_{1o}$ is the resistance facing $C_{\pi 1}$ with all other capacitors open. From Fig. 1.3b,

$$R_{1o} = (R_s + r_x) \parallel r_\pi = 100 \parallel 400 = 80 \text{ ohms} \qquad (1.44)$$

$$\therefore \tau_{1o} = R_{1o}C_{\pi 1} = 0.63 \times 10^{-8} \text{ sec} \qquad (1.45)$$

The calculations for $R_{2o}$, the resistance facing capacitor $C_{\mu 1}$ in the first stage is a little more difficult. If we inject a current at the terminals of $C_{\mu 1}$ and calculate the resulting voltage, with *all* capacitors set to zero, we find:

$$\frac{V}{I} = R_{2o} = R_{1o} + g_m R_{L1} R_{1o} + R_{L1} \qquad (1.46)$$

Using the values from Fig. 1.2, and Eqs. 1.9 and 1.44,

$$R_{2o} = 4.9 \text{ k}$$

Thus,

$$\tau_{2o} = C_{\mu 1} R_{2o} = 1.23 \times 10^{-8} \text{ sec} \qquad (1.47)$$

Again from Fig. 1.3$b$, the resistance $R_{3o}$ facing $C_{\pi 2}$ is:

$$R_{3o} = (R_i + r_x) \parallel r_\pi = 830 \parallel 400 = 270 \text{ ohms} \qquad (1.48)$$

(The resistance $R_{5o}$ facing $C_{\pi 3}$ also equals 270 ohms.) It follows that:

$$\tau_{3o} = C_{\pi 2} R_{3o} = 2.12 \times 10^{-8} \text{ sec} \qquad (1.49)$$

$$\tau_{4o} = C_{\mu 2}(R_{3o} + g_m R_{L2} R_{3o} + R_{L2}) = 4 \times 10^{-8} \text{ sec} \qquad (1.50)$$

$$\tau_{5o} = C_{\pi 3} R_{5o} = 2.12 \times 10^{-8} \text{ sec} \qquad (1.51)$$

$$\tau_{6o} = C_{\mu 3}(R_{5o} + g_m R R_{5o} + R) = 0.75 \times 10^{-8} \text{ sec} \qquad (1.52)$$

Thus

$$\frac{1}{\omega_h} \cong \sum \tau_{jo} = 1.08 \times 10^{-7} \text{ sec} \qquad (1.53)$$

$$\omega_h \cong 9.3 \times 10^6 \text{ rad/sec} \qquad (1.54)$$

or a frequency of 1.5 mc. A more exact calculation indicates that the true 0.707 frequency for this amplifier is 1.8 mc. Thus our approximate value of 1.5 mc is about 16% low.

## 1.3 FINDING APPROXIMATE POLE LOCATIONS

We say in ECP, Chapter 6, that the poles of a network can be related in an approximate manner to the ratios of adjacent coefficients in the denominator polynomial. By way of review, consider the following polynomial:

$$P(s) = (1 - s/s_a)(1 - s/s_b)(1 - s/s_c) \equiv a_0 + a_1 s + a_2 s^2 + a_3 s^3 \qquad (1.55)$$

where in this case

$$a_0 = 1 \quad (1.56a)$$

$$a_1 = -\left(\frac{1}{s_a} + \frac{1}{s_b} + \frac{1}{s_c}\right) \quad (1.56b)$$

$$a_2 = \frac{1}{s_a s_b} + \frac{1}{s_a s_c} + \frac{1}{s_b s_c} \quad (1.56c)$$

$$a_3 = -\frac{1}{s_a s_b s_c} \quad (1.56d)$$

$$\frac{a_2}{a_3} = -(s_a + s_b + s_c) \quad (1.56e)$$

If the poles are widely separated so that $s_a \ll s_b \ll s_c$, we see from Eqs. 1.56a and b that:

$$s_a \cong -\frac{a_0}{a_1} \quad (1.57)$$

and from Eq. 1.56e:

$$s_c \cong -\frac{a_2}{a_3} \quad (1.58)$$

and since $s_a s_b s_c = -1/a_3$, we have

$$s_b \cong -\frac{a_1}{a_2} \quad (1.59a)$$

The results can again be extended to networks with more than three natural frequencies. We can say in general that if the poles are widely separated, the approximate pole locations are given by

$$s_j \cong -\frac{a_{j-1}}{a_j} \quad (1.59b)$$

To indicate the accuracy of this kind of pole calculation, note that for three identical poles at $s = -1$, Eq. 1.59b applied to the coefficients of the cubic equation $(s + 1)^3 = 0$ would yield poles at $s = -\frac{1}{3}, -1$, and $-3$. We conclude that if Eq. 1.59b predicts poles within a factor of three of each other, the calculations are likely to be substantially in error.

Comparison of Eqs. 1.56a and b, and 1.39 with $n = 3$ shows that:

$$-\left(\frac{1}{s_a} + \frac{1}{s_b} + \frac{1}{s_c}\right) = R_{1o}C_1 + R_{2o}C_2 + R_{3o}C_3 \quad (1.60a)$$

Or, more generally,

$$-\sum \frac{1}{s_j} = \sum \tau_{jo} \quad (1.60b)$$

That is, the sum of the open-circuit time constants is *exactly* equal to the negative of the sum of the reciprocals of the natural frequencies of the network.

If one natural frequency is much *smaller* numerically than all the others, as is often the case in multistage amplifiers as a result of interaction between stages, Eq. 1.60 can be approximated by:

$$-\frac{1}{s_l} \cong \sum \tau_{jo} \quad (1.61)$$

where $s_l$ is the lowest natural frequency of the network. Thus *the lowest natural frequency of the network can be estimated simply by forming the sum of the open-circuit time constants* obtained by examining each capacitor with the others open-circuited.

Comparison of Eqs. 1.56e and 1.41 with $n = 3$ shows that:

$$-(s_a + s_b + s_c) = \frac{g_{11}}{C_1} + \frac{g_{22}}{C_2} + \frac{g_{33}}{C_3} \quad (1.62a)$$

Or, more generally,

$$-\sum s_j = \sum \frac{1}{\tau_{js}} \quad (1.62b)$$

That is, the sum of the reciprocal short-circuit time constants is equal to the negative of the sum of the natural frequencies of the network.

If one natural frequency is much *larger* than all the others, Eq. 1.62 can be approximated by:

$$-s_h \cong \sum \frac{1}{\tau_{js}} \quad (1.63)$$

where $s_h$ is the highest natural frequency of the network. That is, the *highest natural frequency of the network can be estimated simply*

## Sec. 1.3 Finding Approximate Pole Locations

by forming the sum of the reciprocals of the short-circuit time constants obtained by examining each capacitor with the others short-circuited.

In terms of the circuit of Fig. 1.3b, if we assume that the lowest pole is much below the rest, then from Eqs. 1.61 and 1.53 the approximate location of the lowest natural frequency is:

$$s_l \cong -\frac{1}{\sum \tau_{jo}} = -0.93 \times 10^7 \text{ sec}^{-1} \quad (1.64)$$

A more accurate calculation shows that the lowest pole is actually located at $-1.18 \times 10^7$ sec$^{-1}$, so our approximate answer is about 21% low.

Note that by "lowest natural frequency" we mean here the lowest natural frequency of the network in Fig. 1.3b. This is clearly not the lowest natural frequency of the complete circuit of Fig. 1.1, because in deriving Fig. 1.3b we short-circuited the coupling and bypass capacitors, thereby eliminating seven low-frequency poles of the transfer function.

If we apply Eqs. 1.61 and 1.63 to the complete amplifier (Fig. 1.1), including coupling capacitors, bypass capacitors, and $C_\pi$ and $C_\mu$ for each stage, we obtain the highest and lowest poles of the *complete* amplifier. The location of these extreme poles may be of considerable importance in feedback amplifier calculations, but often we are more interested in the poles which are dominant in the region of the upper and lower 0.707 frequencies. To find these dominant poles, we must apply Eqs. 1.61 and 1.63 to the *high-frequency* and *low-frequency* models *separately*, as, for example, we did in calculating Eq. 1.64. Equivalently, we can apply Eqs. 1.61 and 1.63 to the high-frequency denominator factors of $A_v(s)$ in Eq. 1.1, and then to the low-frequency factors.

In summary, the methods of this section will give us the approximate lowest and highest poles for *whatever group of poles we are looking at*. Thus, by applying the method to the low-frequency and high-frequency circuits separately, we can find the approximate locations of *the lowest and highest of the low-frequency poles, and the lowest and highest of the high-frequency poles*.

Not only did we use this procedure above in calculating the lowest of the high-frequency poles (Eq. 1.64) but we in effect used it in ECP, Chapter 6, to calculate the low-frequency poles arising from the coupling and bypass capacitors. The reader is urged to review the calculations of pole locations in that chapter in the light of the somewhat more general methods discussed in this chapter.

We continue now by calculating the highest of the high frequency poles from the circuit of Fig. 1.3b, using Eq. 1.63. To do this, we must calculate the *short-circuit* time constants. By inspection of the figure,

$$\frac{1}{\tau_{1s}} = \frac{1/(R_s + r_x) + g_\pi + g_m + G_i + g_x}{C_{\pi 1}} = 2.97 \times 10^9 \text{ sec}^{-1} \quad (1.65)$$

$$\frac{1}{\tau_{2s}} = \frac{G_i + g_x}{C_{\mu 1}} = 8.5 \times 10^9 \text{ sec}^{-1} \quad (1.66)$$

$$\frac{1}{\tau_{3s}} = \frac{g_\pi + g_m + G_i + 2g_x}{C_{\pi 2}} = 3.1 \times 10^9 \text{ sec}^{-1} \quad (1.67)$$

$$\frac{1}{\tau_{4s}} = \frac{G_i + g_x}{C_{\mu 2}} = 8.5 \times 10^9 \text{ sec}^{-1} \quad (1.68)$$

$$\frac{1}{\tau_{5s}} \cong \frac{g_\pi + g_x + g_m + G}{C_{\pi 3}} = 3.1 \times 10^9 \text{ sec}^{-1} \quad (1.69)$$

$$\frac{1}{\tau_{6s}} \cong \frac{G}{C_{\mu 3}} = 8.0 \times 10^9 \text{ sec}^{-1} \quad (1.70)$$

Now from Eq. 1.63, assuming that the highest pole is well separated from the rest (an unjustified assumption, as we shall shortly see):

$$s_h \cong -\sum \frac{1}{\tau_{js}} = -3.42 \times 10^{10} \text{ sec}^{-1} \quad (1.71)$$

a value much in excess of $\omega_T(\cong 3 \times 10^9 \text{ rad/sec})$, as might be expected (see ECP, Chapter 6).

## 1.4 USE OF THE $C_t$ APPROXIMATION TO FIND $\omega_h$ AND THE LOWEST POLE

### 1.4.1 Definition of the $C_t$ Approximation

The preceding development can be used as the basis for another approximate method of analyzing multistage amplifiers. In ECP,

### Sec. 1.4 Use of the $C_t$ Approximation to Find $\omega_h$

Chapter 3, it was shown that for calculating forward gain of a single-stage amplifier with a *resistive load*, we could form an equivalent input circuit as shown in Fig. 1.5a. The time constant associated with capacitor $C_t$ is

$$\tau = R_t C_t = R_t[C_\pi + C_\mu(1 + g_m R_L)] \tag{1.72}$$

where $R_t$ is the total resistance seen by capacitor $C_t$.

Strictly speaking, this kind of calculation cannot be applied to the circuit of Fig. 1.3b, because the loads on the first and second stages are not resistive. However, the calculations of the open-circuit time constants, Eqs. 1.43 and 1.47, for example, show a marked similarity to Eq. 1.72. Specifically,

$$\tau_{1o} + \tau_{2o} = R_{1o}[C_{\pi 1} + C_{\mu 1}(1 + g_m R_{L1})] + C_{\mu 1} R_{L1} \tag{1.73}$$

For most amplifiers, the $C_\mu R_L$ term at the extreme right of Eq. 1.73 will be a negligible part of the open-circuit time constant (because $g_m R_{1o} \gg 1$). Thus, by noting that the definition for $R_t$ is consistent with that for $R_{1o}$, we find:

$$\tau_{1o} + \tau_{2o} \cong R_{1o} C_{t1} = R_{t1} C_{t1} \tag{1.74}$$

where $C_{t1}$ is calculated by assuming a *resistive load* on the first transistor, that is, assuming $C_{\pi 2}$ and $C_{\mu 2}$ are open-circuited. For ease of reference we shall refer to this approximate calculation of $C_t$ as the $C_t$ *approximation*. It follows from Eq. 1.74 that we can obtain a good estimate of $\sum \tau_{jo}$ for the whole amplifier just by summing the three time constants $R_t C_t$ obtained by using this $C_t$ approximation. That is,

$$\sum \tau_{jo} \cong \sum R_t C_t \tag{1.75}$$

This method of calculating $\sum \tau_{jo}$, and hence estimating the 0.707 point and the lowest natural frequency is particularly convenient because the time constants $R_t C_t$ are so easy to find by inspection.

The $C_t$ approximation can also be made plausible in terms of the circuit. We are looking for the frequency $\omega_h$ where the magnitude of the output voltage drops to 0.707 of its mid-frequency value. In a circuit with many capacitors, often at $\omega_h$ each $RC$ circuit will contribute only a small amount to this reduction in gain. Under these conditions, it is reasonable to calculate the effects of the different $RC$ circuits one at a time, by assuming that the other $RC$

## 24 Gain and Bandwidth Calculations: A First Approximation

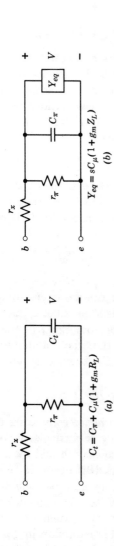

**Fig. 1.5.** Single-stage amplifier approximate input circuits for calculating $V$. (a) For resistive load $R_L$ only. (b) For any load $Z_L$.

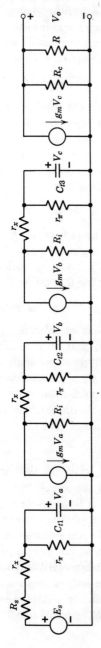

**Fig. 1.6.** Approximate circuit model based on $C_t$ approximation. (Good only for forward gain and input impedance calculations.)

circuits are essentially contributing nothing to the decrease in gain, that is, the other capacitors are open-circuited. Essentially we are treating the 0.707 point calculation as a first-order perturbation on the mid-band calculation, by computing the effect of each capacitor singly, and ignoring interaction.

### 1.4.2 Example

To illustrate the use of the $C_t$ approximation, we again calculate the 0.707 frequency of the amplifier in Fig. 1.3b. It is sometimes convenient for this purpose to put the $C_t$ approximation in the form of an approximate circuit model, as shown in Fig. 1.6. However, there are some very important restrictions on the use of this model. Specifically, it can be used to calculate forward gain and input impedance up to a little above the half-power frequency $\omega_h$, *but cannot ever be used to calculate reverse gain or output impedance.* This restriction will not be troublesome in this chapter, because we are interested mainly in forward gain and bandwidth calculations.

From previous calculations, Sec. 1.1 and 1.2.3,

$$R_{t1} = R_{1o} = 80 \ \Omega \qquad R_{L1} = 285 \ \Omega$$

$$R_{t2} = R_{3o} = 270 \ \Omega \qquad R_{L2} = 285 \ \Omega$$

$$R_{t3} = R_{5o} = 270 \ \Omega \qquad R_{L3} = 50 \ \Omega$$

Thus,

$$R_{t1}C_{t1} = R_{t1}[C_\pi + C_\mu(1 + g_m R_{L1})]$$

$$= 80[78.5 + 2.5(1 + 0.2 \times 285)] \times 10^{-12}$$

$$= 1.8 \times 10^{-8} \text{ sec} \tag{1.76}$$

Similarly,

$$R_{t2}C_{t2} = 6 \times 10^{-8} \text{ sec} \tag{1.77}$$

$$R_{t3}C_{t3} = 2.9 \times 10^{-8} \text{ sec} \tag{1.78}$$

From Eqs. 1.42 and 1.75:

$$\omega_h \cong \frac{1}{\sum R_t C_t} = \frac{10^8}{10.7} = 9.4 \times 10^6 \text{ rad/sec} \tag{1.79}$$

which checks closely with the approximate value calculated in Sec. 1.2.3, but is about 16% below the true value. Also, from Eqs. 1.61 and 1.75 we find the approximate location of the lowest pole to be:

$$s_1 \cong \frac{-1}{\sum R_t C_t} = -9.4 \times 10^6 \text{ sec}^{-1} \qquad (1.80)$$

## 1.5 EXTENSION OF THE $C_t$ APPROXIMATION TO FIND $A_v(j\omega)$

It can be seen from Eqs. 1.75 and 1.39 that the $C_t$ approximation in effect gives an estimate of the first two terms of the denominator polynomial of $A_v$:

$$A_v = \frac{A_{vo}}{1 + s \sum R_t C_t + \ldots} \qquad (1.81)$$

We have already used this fact to calculate the approximate 0.707 frequency and lowest pole. A useful extension of the $C_t$ approximation is based on Eqs. 1.75 and 1.60b. That is,

$$\sum R_t C_t \cong -\sum \frac{1}{s_j} \qquad (1.82)$$

where the right-hand side represents the sum of the reciprocals of the *true* network natural frequencies. If we arbitrarily define three "substitute" natural frequencies:

$$s_x = \frac{-1}{R_{t1} C_{t1}} \qquad (1.83a)$$

$$s_y = \frac{-1}{R_{t2} C_{t2}} \qquad (1.83b)$$

$$s_z = \frac{-1}{R_{t3} C_{t3}} \qquad (1.83c)$$

then from Eq. 1.82:

$$\frac{1}{s_x} + \frac{1}{s_y} + \frac{1}{s_z} \cong \sum \frac{1}{s_j} \qquad (1.84)$$

### Sec. 1.5 Extension of the $C_t$ Approximation to Find $A_v(j\omega)$

where $s_j$ again indicates the *true* natural frequencies of the network. The substitute poles of $A_v$ as defined by Eqs. 1.83 can now be used to calculate the magnitude and phase of $A_v(j\omega)$. Two things should be remembered, however, when applying the $C_t$ approximation in this manner:

1. Because we are really using only the first two terms of the denominator polynomial, the calculation of $A_v(j\omega)$ will have some error at $\omega_h$, but should be acceptable up to at least that frequency.

2. The substitute pole locations defined by Eqs. 1.83 differ from the true pole locations by substantial amounts. They are invariably bunched more closely together because in effect we have ignored the interaction among the capacitors. The lowest substitute pole is often higher than the true lowest pole by about 20 or 30%, and the highest substitute pole may be lower than it should be by as much as an order of magnitude.

Nevertheless, because the substitute poles are related to the real poles by Eq. 1.84, and because of the relative insensitivity of $|A_v(j\omega)|$ to the precise location of the poles on the real axis, the amplitude response can be calculated with fair accuracy up to and beyond $\omega_h$ by using these substitute poles.

For our three-stage amplifier example we obtain from Eqs. 1.83:

$$s_x = -1.67 \times 10^7 \text{ sec}^{-1}$$

$$s_y = -3.5 \times 10^7 \text{ sec}^{-1}$$

$$s_z = -5.6 \times 10^7 \text{ sec}^{-1}$$

Thus,

$$A_v \cong \frac{K}{(s + 0.0167)(s + 0.035)(s + 0.056)} \quad (1.85)$$
$$(s \text{ in units of nanosec}^{-1})$$

These numbers should be compared with the more accurate calculations given in Sec. 2.3.1.

The magnitude and phase of $A_v(j\omega)$ can now be evaluated by a simple extension of the method discussed in ECP, Sec. 3.1.4, for plotting single-pole functions. Specifically, we can plot the magnitude of each of the factors in Eq. 1.85 versus $\omega$ on log-log coordinates, then add the logarithmic plots to get the over-all $|A_v(j\omega)|$.

## 28   Gain and Bandwidth Calculations: A First Approximation

Recall that the magnitude of each of the factors can be approximated by its two asymptotes, one of slope zero, the other of slope $-1$, intersecting at $\omega = |s_o|$, where $s_o$ is the pole frequency. We can thus draw a composite "asymptote" by drawing the asymptotes for each factor, then summing. However, it is faster to form the resultant "asymptote" directly as shown in Fig. 1.7. The procedure is as follows. Draw the mid-frequency asymptote with zero slope at $A_v = A_{v0}$. Draw the second asymptote with slope $-1$ intersecting the first at $\omega = 0.0167 \times 10^9$. Draw the third asymptote with slope $-2$, intersecting the second at $\omega = 0.035 \times 10^9$, and so on. If there had been any zeros of $A_v(s)$, these factors would have been handled in the same way, except that the slope would be *increased* by one integer, instead of decreased. To approximate the phase of $A_v(j\omega)$, we recall that for a function with a single pole at $s_o$, the phase is $-45°$ at $\omega = |s_o|$, $-6°$ at $\omega = |s_o|/10$, and $-84°$ at $\omega = 10|s_o|$. It is fairly easy to sketch the approximate phase of $A_v(j\omega)$ by summing the individual phase plots based on the above values. This approximate method of sketching the magnitude and phase of a function from its factored polynomials is often called a *Bode plot*.

A more exact solution for $A_v(j\omega)$, obtained by solving the six-by-six determinant corresponding to Fig. 1.3$b$, is also shown in the figure. Note that our calculation of $|A_v(j\omega)|$ from Eq. 1.85 is too *large* at $\omega_h (\cong 10^7$ rad/sec here). This will be true in general for this approximation (because the substitute poles of the $C_t$ approximation are closer together than the actual poles of the network, but we used nearly identical values of $a_0$ and $a_1$ in the two gain expressions). Also, the approximate $|A_v(j\omega)|$ is too *small* at high frequencies, because the load impedance on the first and second stages actually approaches $r_x || R_i$ at high frequencies, whereas the $C_t$ calculation assumes the load on these stages is always $R_L = (r_x + r_\pi) || R_i$. Because our approximate calculation of $|A_v(j\omega)|$ must therefore cross the more exact solution at a frequency above $\omega_h$, we can see from Fig. 1.7 that this approximate analysis gives acceptable results out to frequencies well beyond $\omega_h$. We therefore somewhat arbitrarily take the figure $3\omega_h$ as a rough upper limit on the range of validity of this representation. The plots based on Eq. 1.85 are shown dotted above $3 \times 10^7$ rad/sec to emphasize this limit.

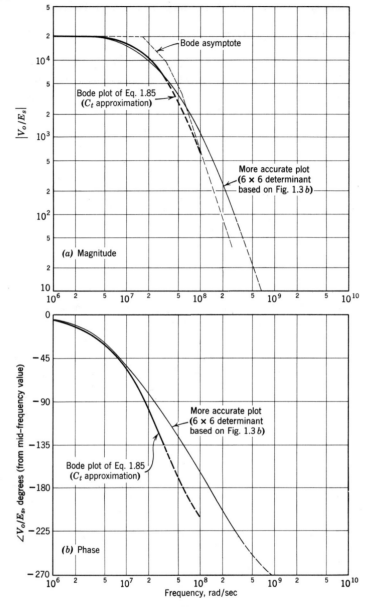

**Fig. 1.7.** Over-all amplifier response. These plots, and subsequent amplifier response plots in this book, do not show the 180° phase shift of the amplifier at mid-frequencies.

## 1.6 BEHAVIOR AT HIGH FREQUENCY

Some approximation to the behavior of the circuit in Fig. 1.3b at high frequencies can be obtained by network inspection techniques similar to those discussed above. Specifically, at frequencies above the transverse cut-off frequency

$$\omega_b = \frac{g_x + g_\pi}{C_\pi + C_\mu} \quad (1.86)$$

the load on each transistor stage again begins to look resistive, because the combined effects of $C_\pi$ and $C_\mu$ give rise to an impedance which is a short-circuit compared to $r_x$. Thus the first-stage load, for example, reduces in this frequency range to $R_i \parallel r_x$. (See Fig. 1.3b.) The circuit in effect breaks up into three isolated stages, which means that there are three high-frequency poles, one associated with each stage. Each of these three high-frequency poles can be found by summing the two short-circuit time constants associated with each stage, as in Sec. 1.3. Inspection of Eqs. 1.65 to 1.70 will show that, in effect, these calculations have already been made, and that:

$$s_4 = -\left(\frac{1}{\tau_{1s}} + \frac{1}{\tau_{2s}}\right) = -11.5 \times 10^9 \text{ sec}^{-1} \quad (1.87a)$$

$$s_5 = -\left(\frac{1}{\tau_{3s}} + \frac{1}{\tau_{4s}}\right) = -11.6 \times 10^9 \text{ sec}^{-1} \quad (1.87b)$$

$$s_6 = -\left(\frac{1}{\tau_{5s}} + \frac{1}{\tau_{6s}}\right) = -11.1 \times 10^9 \text{ sec}^{-1} \quad (1.87c)$$

The apparent conflict between Eqs. 1.87 and 1.71 is readily resolved. Equation 1.71 applies only if $s_6$ is far above the other poles, and clearly here it is not. In fact, the three poles $s_4$, $s_5$, and $s_6$ are very close together, and this group is widely separated from $s_1$, $s_2$, and $s_3$. Thus the correct approximation to Eq. 1.62b is not Eq. 1.63, but rather

$$s_4 + s_5 + s_6 \cong -\sum \frac{1}{\tau_{js}} \quad (1.88)$$

This is consistent with Eqs. 1.87.

Equations 1.87 do *not* indicate that the *amplifier* of Fig. 1.1 actually has natural frequencies in the vicinity of $-11 \times 10^9$ sec$^{-1}$, because the model we are using, the intrinsic hybrid-$\pi$, is not valid for such a calculation. It is correct to say, however, that within the frequency range where our intrinsic hybrid-$\pi$ model is valid, the amplifier behaves *as if* there were poles in this general region.

## PROBLEMS

**P1.1** The circuit of a three-stage RC coupled transistor amplifier is shown in Fig. 1.8. The element values are: $R_1 = 15$ k, $R_2 = 5$ k, $R_4 = 1$ k, $R_3 = 1.5$ k, $\beta_F = 50$, $R_s = R_L = 100$ ohms.

(a) Estimate the operating point ($I_C$, $V_{CE}$) of the transistors (all three stages have the same bias network). *Suggestion:* Assume that $I_B$ is much less than the current in $R_1$ and $R_2$ and assume that $V_{BE} = 0.6$ volt. Are these assumptions justified for a silicon transistor?

(b) Construct a complete incremental model for the amplifier, showing numerical values for all elements in the model. Assume that the three transistors are identical and have the following properties: $\beta_0 = g_m r_\pi = 50$, $C_\mu = 3$ pf, $f_T = 250$ mc, and $r_x = 50$ ohms. Neglect $r_o$ and $r_\mu$. Is this reasonable? Also assume that all coupling and bypass capacitors have negligible ac voltage drops across them.

(c) What is the mid-band voltage gain $V_o/V_i$ of the amplifier? What is the mid-band input resistance $V_i/I_i$? What is the mid-band value of the over-all voltage gain $V_o/E_s$?

(d) Estimate the upper half-power frequency of this amplifier.

**P1.2** For the amplifier shown in Fig. 1.8, assume that the resistors $R_1$, $R_2$, $R_3$, and $R_4$ have been adjusted so that each transistor operates at $I_C = 10$ ma. At this operating point the transistor parameters are $\beta_0 = 100$, $r_x = 50$ ohms, $f_T = 160$ mc, $C_\mu = 3$pf.

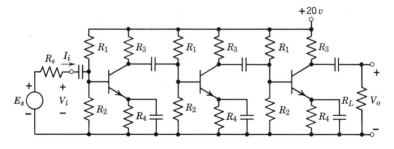

Fig. 1.8

**32** *Gain and Bandwidth Calculations: A First Approximation*

    (a) Find the incremental model of the amplifier, assuming signal voltage drops across coupling and bypass capacitors are negligible and the biasing resistors $R_1$ and $R_2$ can be neglected in comparison with $r_x$ and $r_\pi$.

    (b) Calculate the mid-frequency gain of this amplifier, assuming $R_3 = R_s = R_L = 300$ ohms.

    (c) Estimate the high-frequency 0.707 frequency $\omega_h$ of the amplifier.

    (d) On the basis of the $C_t$ approximation, plot $|A_v(j\omega)|$ versus $\omega$ on log-log coordinates. What is the range of validity of this approximate frequency response?

**P1.3** The small-signal model of a two-stage transistor amplifier driven from a low-impedance source is shown in Fig. 1.9.

    (a) Determine the low-frequency voltage gain $V_o/V_s$.

    (b) Determine the 0.707 frequency $\omega_h$ of the amplifier. If you use an approximate method, be sure to state your approximations clearly.

**P1.4** If the gain of an amplifier is given by

$$A_v = \frac{(1+10s)(1+s)}{(1+4s)(1+.1s)}$$

    (a) Plot the asymptotic curves of $|A_v|$ and $\angle A_v$ versus $\omega$.

    (b) On the basis of the asymptotes in (a), make a free-hand sketch of $|A_v|$ and $\angle A_v$ versus $\omega$ without recourse to any calculations.

    (c) Calculate the exact gain at enough points to improve your sketch. Evaluate any large errors in your initial sketch.

**P1.5** Repeat P1.4 for a gain given by

$$A_v = \frac{(1+10s)(1-s)}{(1+2s)(1+0.1s)^2}$$

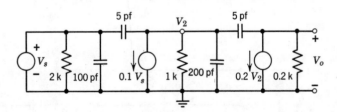

Fig. 1.9

# 2

## Amplifier Calculations Using a π Model

### 2.0 INTRODUCTION

In the preceding chapter we considered methods of finding the approximate upper 0.707 point, approximate locations of the lowest pole of a multistage amplifier, and the gain expression valid up to $3\omega_h$. Often, however, more complete information concerning the frequency response or pole locations is needed; for example, in feedback amplifier design. One analysis method good up to $\omega_T[=g_m/(C_\pi + C_\mu)]$ would be to represent each amplifier stage by the hybrid-$\pi$ model, complete with extrinsic elements, Fig. 2.1 (a duplicate of Fig. 3.12, ECP), and solve the node equations for the resulting network. However, even for the simple three-stage amplifier of Fig. 1.1, this procedure is laborious, because we are forced to solve a seven-by-seven determinant.

Analysis via node equations is certainly possible, regardless of the complexity of the circuit and, in fact, a direct solution of simultaneous equations may be the easiest course if detailed results are required. Often, however, we do not need this much detail. In particular, calculations of the frequency response above $\omega_T$ are of no value. In fact, because the gain of an iterative low-pass amplifier is always less than one at $\omega_T$, calculations up to, say, $\omega_T/10$ are

## 34   Amplifier Calculations Using a π Model

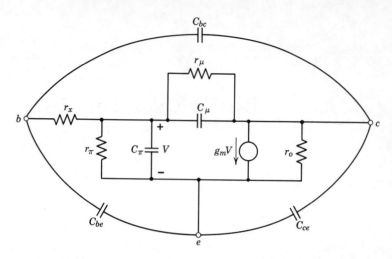

**Fig. 2.1.** Hybrid-$\pi$ model with capacitors added to represent header capacitance and overlap-diode capacitance.

often entirely sufficient. This fact is of considerable importance. As we saw in Chapter 3 of ECP, transistor models can be simplified considerably, with a resulting reduction in computational labor, if we are willing to restrict the frequency range over which the analysis is valid.

Considerable simplification of calculations can also be obtained when node equations are used if we can manage to absorb $r_x$ into the rest of the model, thereby converting the "hybrid-$\pi$" into a "true-$\pi$" topology. To this end, in this chapter we develop $\pi$ models which are good up to $\omega_b/3$, where $\omega_b$, the transverse cut-off frequency, is:

$$\omega_b = \frac{g_x + g_\pi}{C_\pi + C_\mu} \qquad (2.1)$$

This frequency range is more limited than the range of either the intrinsic hybrid-$\pi$ model ($\omega_b$), or the hybrid-$\pi$ with extrinsic elements ($\omega_T$). However, the resulting models will be better suited topologically for node equation analysis.

## 2.1 TRANSISTOR MODEL SIMPLIFICATION BY ABSORBING $r_x$

### 2.1.1 Model Derived from y-Parameter Data

The $y$-parameter data shown in Fig. 2.2 (a duplicate of Fig. 3.16, ECP) shows clearly that below about 6 mc, the $y$-parameters for this particular transistor can be represented by simple $RC$ networks and a dependent source. Specifically, below 6 mc $y_{ie}$ can be approximated by a parallel $RC$, $y_{fe}$ by a constant $g_m$, $y_{re}$ by a single capacitor, and $y_{oe}$ by a parallel $RC$. Such a model is shown in Fig. 2.3. The numerical values were obtained by calculating slopes and intercepts from the curves in Fig. 2.2.

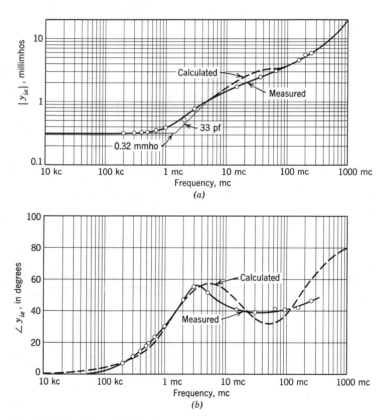

**Fig. 2.2.** Experimental data of $y$-parameters versus frequency. Results calculated from hybrid-$\pi$ model of Fig. 2.1 show as dashed lines for comparison.

## 36 Amplifier Calculations Using a π Model

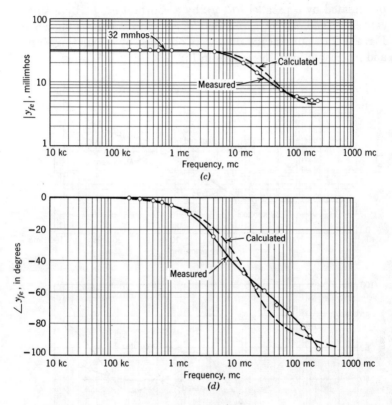

**Fig. 2.2** (*Continued*)

## Sec. 2.1 Transistor Model Simplification by Absorbing $r_x$

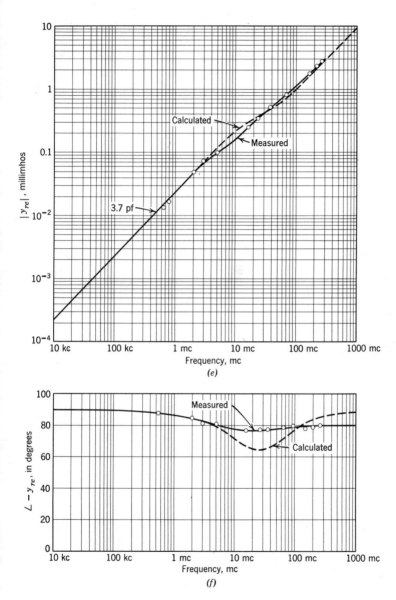

**Fig. 2.2** (*Continued*)

## 38  Amplifier Calculations Using a π Model

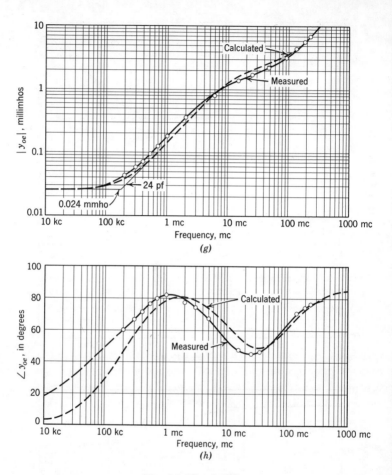

**Fig. 2.2** (*Concluded*)

We have chosen 6 mc as the limit of validity for this representation on the basis that this is about one third of the break frequency $\omega_b$ for all of the curves. A simple calculation will show that we can completely neglect the effects of the pole at $s = -\omega_b$ for all frequencies below $\omega_b/3$, and introduce at most an error of 6% in amplitude and 19° in phase.

## Sec. 2.1 Transistor Model Simplification by Absorbing $r_x$

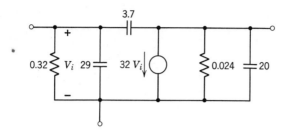

$$[Y] \cong \begin{bmatrix} 0.32 + s\,33 & -s\,3.7 \\ 32 & 0.024 + s\,24 \end{bmatrix}$$

**Fig. 2.3.** Circuit equivalent of $y$-parameter data valid up to 6 mc. (Units of millimhos and picofarads.) Values were obtained by calculating slopes and intercepts of the $y$-parameter data in Fig. 2.2.

### 2.1.2 Relation to Hybrid-$\pi$ Model

To find the relationship between the model parameters in Fig. 2.3 and the hybrid-$\pi$ parameters, we write the $y$-parameters in matrix form for the intrinsic hybrid-$\pi$ model, Fig. 2.4a. Note that we are tacitly assuming that $g_\mu$ and $g_o$ are negligible.

$$[Y] = \begin{bmatrix} \dfrac{[g_\pi + s(C_\pi + C_\mu)]g_x}{g_x + g_\pi + s(C_\pi + C_\mu)} & \dfrac{-sC_\mu g_x}{g_x + g_\pi + s(C_\pi + C_\mu)} \\ \dfrac{(g_m - sC_\mu)g_x}{g_x + g_\pi + s(C_\pi + C_\mu)} & \dfrac{sC_\mu(g_x + g_\pi + sC_\pi) + sC_\mu g_m}{g_x + g_\pi + s(C_\pi + C_\mu)} \end{bmatrix} \quad (2.2)$$

It is clear from these equations why the hybrid-$\pi$ model is not ideal for nodal analysis! If, as before, we restrict the range of validity of the representation to below 6 mc, or more generally, below $\omega_b/3$, then

$$g_x + g_\pi + s(C_\pi + C_\mu) \cong g_x + g_\pi \quad (2.3)$$

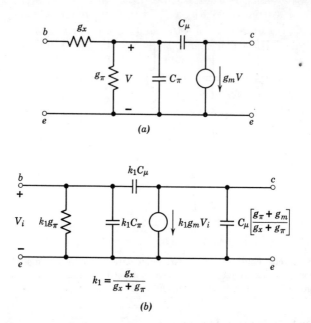

**Fig. 2.4.** Transformation from the intrinsic hybrid-$\pi$ model to a $\pi$ model.

It follows that, for $\omega < (\omega_b/3)$,

$$[Y] = \begin{bmatrix} k_1[g_\pi + s(C_\pi + C_\mu)] & -k_1 s C_\mu \\ k_1(g_m - sC_\mu) & k_1 s C_\mu + \left(\dfrac{g_\pi + g_m}{g_x + g_\pi}\right) s C_\mu \end{bmatrix} \quad (2.4)$$

where $k_1 = \dfrac{g_x}{g_x + g_\pi}$.

Figure 2.4b is an interpretation in terms of a circuit model of the $y$-parameters in Eq. 2.4, a model identical in form with that of Fig. 2.3. In effect $r_x$ has been absorbed into the model by changing $g_\pi$, $C_\pi$, $C_\mu$, and $g_m$ by the factor $k_1$ (typically between 0.8 and 0.99). In addition, it is necessary to add a capacitance from collector to emitter of value $C_\mu(g_\pi + g_m)/(g_x + g_\pi)$. The factor multiplying $C_\mu$ here will lie typically between 1.5 and 50.

One way of visualizing the origin of the capacitor from collector to emitter in Fig. 2.4b is to note that the effective capacitance $C_{oe}$ from the collector node to ground is increased, because in the hybrid-$\pi$ model with base shorted to emitter, for every unit of current through $C_\mu$, the dependent source generates $g_m/(g_x + g_\pi)$ units of current. Hence, at low frequencies, the effective short-circuit output capacitance is:

$$C_{oe} = \left(1 + \frac{g_m}{g_x + g_\pi}\right) C_\mu = k_1 C_\mu + \left(\frac{g_\pi + g_m}{g_x + g_\pi}\right) C_\mu \quad (2.5)$$

Note that, by absorbing $r_x$, we have derived a model much more amenable to nodal analysis. Another important advantage of this model is that extrinsic overlap-diode and header capacitance can be added without in any way complicating the model configuration or analysis. In fact, the model of Fig. 2.3, because it was derived directly from measured $y$-parameter data, already includes these extrinsic elements.

## 2.2 ILLUSTRATIVE EXAMPLE: THREE-STAGE AMPLIFIER

### 2.2.1 Model Parameters

To illustrate some of these ideas, let us calculate the transfer function of the three-stage amplifier discussed in Sec. 1.1. The transistor we have been studying thus far in this chapter is actually very similar to the one used in the amplifier of Sec. 1.1, when we note that in the amplifier, $I_C = 5$ ma, whereas in the $y$-parameter curves and resulting model discussed in Sec. 2.1, $I_C$ is slightly less than 1 milliampere. However, *to permit direct intercomparison of the approximate methods of Chapter 1 with those in this chapter*, we modify the parameters in Fig. 2.3 to correspond exactly with the values used in Chapter 1. Thus $g_m = 200$ mmho, $g_\pi = 2.5$ mmhos, $g_x = 20$ mmhos, $C_\pi = 78.5$ pf. The parameter $C_\mu = 2.5$ pf in Chapter 1 really represents the combined effects of the intrinsic $C_\mu$ and the overlap diode and header capacitance (see Chapter 3, ECP). We assume, therefore, that the true intrinsic $C_\mu = 1.5$ pf, and $C_{bc} = 1$ pf. In addition, we assume $C_{be} = 0.5$ pf, $C_{ce} = 0.5$ pf.

To find the parameters of the model in Fig. 2.4, we apply the relations shown in the figure or Eq. 2.4, and then add the extrinsic capacitance. The resulting values are given in Fig. 2.5. In accordance with Eq. 2.4, this model is good to about $\omega_b/3$, where in this example:

$$\omega_b = \frac{g_x + g_\pi}{C_\pi + C_\mu} = \frac{22.5 \text{ mmhos}}{80 \text{ pf}} = 0.28 \times 10^9 \text{ rad/sec}$$

That is, the model is good to about $10^8$ rad/sec, or 15 mc.

### 2.2.2 Finding Transfer Function by Nodal Analysis

Using this model for the transistors, we obtain the circuit model for the three-stage amplifier as shown in Fig. 2.6. This model is good up to about 15 mc. The 1.3 mmho interstage conductances are the elements called $G_i$ (or $R_i$) in the preceding chapter, that is, $G_i = 1/R_c + 1/R_b$. On the basis of the model, we can write the following $Y$ matrix, with units of millimhos and picofarads:

$$[Y] = \begin{bmatrix} 22.2 + s\,72.3 & -s\,2.3 & & \\ 180 - s\,2.3 & 3.5 + s\,88.6 & -s\,2.3 & \\ & 180 - s\,2.3 & 3.5 + s\,88.6 & -s\,2.3 \\ & & 180 - s\,2.3 & 20 + s\,16.3 \end{bmatrix}$$

(2.6)

Because of the units chosen for $g$ and $C$, the variable $s$ will have the units of mmhos/pf or (nanoseconds)$^{-1}$. Thus to convert back to the usual units of seconds$^{-1}$, we must multiply all values of $s$ calculated from these equations by $10^9$.

In the succeeding calculations we shall neglect the three $sC_\mu$ terms in comparison with $g_m$; that is, we neglect $-s\,2.3$ in the terms $180 - s\,2.3$, because these contribute only to the three poles and three right-half plane zeros much larger in magnitude than $\omega_T$. On this basis, the voltage gain is, by Cramer's rule,

$$A_v(s) = \frac{V_o}{E_s} = \frac{20V_o}{I_i} = -\frac{20(180)(180)(180)}{\Delta}$$

(2.7)

$$= \frac{-1.16 \times 10^4}{(0.54 + s\,54 + s^2\,720 + s^3\,1860 + s^4\,920)}$$
($s$ in units of nanosec $^{-1}$)

## Sec. 2.2 Illustrative Example: Three-Stage Amplifier 43

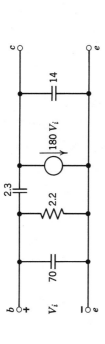

**Fig 2.5.** A $\pi$ model for transistor given in Sec. 1.0. Values are in millimhos and picofarads. Model good to about 15 mc.

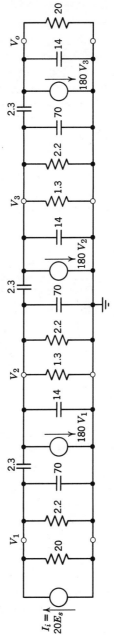

**Fig. 2.6.** Circuit model for 3-stage amplifier. (Units of millimhos and picofarads.)

## 2.2.3 Solution by Signal Flow Graphs

Those familiar with signal flow graphs may prefer to solve the circuit in Fig. 2.6 by this method rather than by determinants. To obtain a set of equations which can readily be interpreted in terms of a flow graph, we divide through each of the node equations represented in Eq. 2.6 by the real part of the term on the principal diagonal. Thus the first node equation becomes, for example,

$$\frac{20}{22.2} E_s = \left(1 + s \frac{72.3}{22.2}\right) V_1 - \frac{s\, 2.3}{22.2} V_2 \qquad (2.8)$$

The flow graph for this slightly modified set of node equations is shown in Fig. 2.7a. Note that, in this form, each loop gain has the form $s\tau$, where the time constants $\tau$ are closely related to those

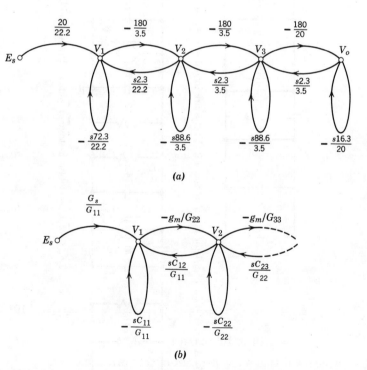

**Fig. 2.7.** Signal flow graphs. (a) Graph for circuit in Fig. 2.6. ($s$ in units of nanoseconds$^{-1}$.) (b) Transistor amplifier graph in literal form.

Sec. 2.3 Performance Calculations Based on Transfer Function

discussed in Chapter 1. Calculation of $V_o/E_s$ by the usual flow graph rules yields Eq. 2.7 as before.

The same flow graph in literal form is shown in Fig. 2.7b. The elements are defined in accordance with the usual matrix and determinant notation as shown below:

$$[Y] = \begin{bmatrix} G_{11} + sC_{11} & -sC_{12} & & \\ g_m & G_{22} + sC_{22} & -sC_{23} & \\ & g_m & G_{33} + sC_{33} & -sC_{34} \\ & & \cdot & \cdot & \cdot \\ & & & \cdot & \cdot & \cdot \end{bmatrix} \quad (2.9)$$

On this basis, the identity of the branches in the flow graph becomes clear. The self-loop at each node has a time constant equal to the total shunt capacitance at that node divided by the total shunt conductance. The forward branch between nodes is $-g_m R_{22}$, the forward mid-frequency voltage gain. The reverse branch between nodes has a time constant equal to the mutual capacitance divided by the total shunt conductance of the input node. Figure 2.7b and Eq. 2.9 provide the necessary information for construction of the flow graph for any network of the form of Fig. 2.6.

## 2.3 PERFORMANCE CALCULATIONS BASED ON TRANSFER FUNCTION

### 2.3.1 *Calculation of Approximate 0.707 Frequency and Approximate Pole Locations*

Using the approximate methods for calculating the upper 0.707 point outlined in Chapter 1, we find from Eq. 2.7 as a first approximation:

$$\omega_h \cong \frac{a_0}{a_1} = \frac{0.54 \times 10^9}{54} = 10^7 \text{ rad/sec}$$

As a second try:

$$\omega_h \cong \frac{a_0}{\sqrt{a_1^2 - 2a_0 a_2}} = \frac{0.54}{\sqrt{(54)^2 - 2(0.54)(720)}} \times 10^9$$

$$= 1.16 \times 10^7 \text{ rad/sec}$$

which is within a few per cent of the correct answer.

In Chapter 1 we were able to make an estimate of the lowest and highest poles of the network. Because we now have a more complete expression for the voltage gain, Eq. 2.7, we can use the method of Sec. 1.3 to find the approximate location of four of the poles of the transfer function $V_o/E_s$:

$$s_1 \cong -\frac{a_0}{a_1} = -\frac{0.54}{54} \times 10^9 = -10^7 \text{ sec}^{-1}$$

$$s_2 \cong -\frac{a_1}{a_2} = -\frac{54}{720} \times 10^9 = -7.5 \times 10^7 \text{ sec}^{-1}$$

$$s_3 \cong -\frac{a_2}{a_3} = -\frac{720}{1860} \times 10^9 = -3.9 \times 10^8 \text{ sec}^{-1}$$

$$s_4 \cong -\frac{a_3}{a_4} = -\frac{1860}{920} \times 10^9 = -2.02 \times 10^9 \text{ sec}^{-1}$$

To check on these approximate poles, we can find the pole locations in two other ways. First, we can directly solve for the roots of the denominator polynomial of Eq. 2.7. Second, we can expand the six-by-six determinant corresponding to the circuit in Fig. 1.3b, and find the roots of the resulting sixth-order polynomial. The results of both of these calculations, together with the approximate pole locations calculated above, are included in Table 2.1. Also included in the table are the results of pole locations calculated by the approximate methods discussed in Secs. 1.3, 1.5, and 1.6. All values are in units of (nanosec)$^{-1}$.

**TABLE 2.1**

| Method<br>Poles* | Ratio of Adjacent Coefficients, Eq. 2.7 | Roots of Eq. 2.7 | Roots of 6th Order Eq. for Fig. 1.3b | Sec. 1.3 Sum of Time Constants, (Eqs. 1.64, 1.71) | Sec. 1.6 Direct Circuit Calcs. (Eqs. 1.87) | Sec. 1.5 Substitute Poles (Eq. 1.85) |
|---|---|---|---|---|---|---|
| $s_1$ | −0.01 | −0.0118 | −0.0110 | −0.0093 | | −0.017 |
| $s_2$ | −0.075 | −0.0835 | −0.0713 | | | −0.035 |
| $s_3$ | −0.39 | −0.389 | −0.187 | | | −0.056 |
| $s_4$ | −2.02 | −1.53 | −10 | | −11.1 | |
| $s_5$ | | | −12 | | −11.5 | |
| $s_6$ | | | −12 | −34.2 | −11.6 | |

* $s$ in units of nanosec$^{-1}$

The location of the two lowest poles $s_1$ and $s_2$ calculated from Eq. 2.7, either by direct factoring or by forming the ratio of the

### Sec. 2.3 Performance Calculations Based on Transfer Function 47

coefficients, agrees well with the calculations from the six-by-six determinant. However, we cannot expect $s_3$ and $s_4$ calculated from Eq. 2.7 to be as accurate because of the limitations of our $\pi$ model. Nonetheless, circuit performance based on Eq. 2.7 and on the approximate pole locations should be quite accurate below $\omega_b/3$, or 15 mc in this example.

When high-frequency poles $s_4$, $s_5$, and $s_6$ are calculated from the six-by-six determinant for Fig. 1.3b, only about 10% accuracy is realized unless the computations are carried out to four significant figures, because these three poles are so close together. As pointed out in Sec. 1.6, the network breaks up at high frequencies into three isolated units, and because the interaction between the three nearly identical units is so small, the three high-frequency poles are very close together. Because of the problem of computational accuracy in the six-by-six determinant under these circumstances, the most accurate estimate of $s_4$, $s_5$, and $s_6$ in Table 2.1 results from inspection of the circuit, in the manner discussed in Sec. 1.6. (Recall, however, that because of fundamental model inaccuracies, none of these values for $s_4$, $s_5$, and $s_6$ is correct for the *actual amplifier* Fig. 1.1. See page 31.)

### 2.3.2 *Calculations of Magnitude and Phase of $A_v$ ($j\omega$) from Factored Polynomial*

There are several ways of calculating the magnitude and phase of $V_o/E_s$ from Eq. 2.7. Perhaps the most familiar method is to factor the denominator polynomial (as we did in the preceding section) and make a Bode plot based on these factors. For convenience, Eq. 2.7 has been rewritten below in factored form:

$$\frac{V_o}{E_s} = \frac{-1.27}{(s + 0.0118)(s + 0.0835)(s + 0.389)(s + 1.53)} \quad (2.10)$$
$$\text{($s$ in units of nanosec}^{-1})$$

The resulting plots of the magnitude and phase of $V_o/E_s$ versus $\omega$ are shown in Fig. 2.8. Because these calculations are based on the $\pi$ model in Fig. 2.5, we have confidence in the result out to $\omega_b/3$, or $10^8$ rad/sec for this example. The curves are shown as dashed lines beyond this frequency to emphasize this point.

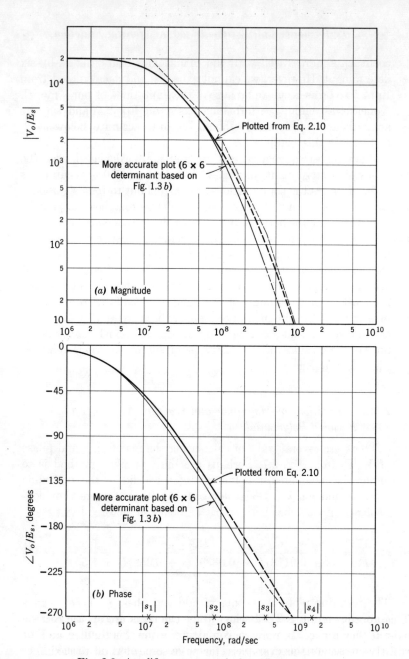

**Fig. 2.8.** Amplifier response calculated from $\pi$ model.

### Sec. 2.3 Performance Calculations Based on Transfer Function

Also shown for comparison in Fig. 2.8 are the magnitude and phase curves calculated from the six-by-six determinant for the circuit in Fig. 1.3b. This calculation, based on the intrinsic hybrid-$\pi$ model, is good out to $\omega_b$, or $3 \times 10^8$ rad/sec. The agreement between this curve and the Bode plot based on Eq. 2.10 is obviously very good. Any discrepancy is within the expected limits based on the approximations involved in the two models used, especially in view of the somewhat arbitrary division of $C_\mu$ and $C_{bc}$ made in Sec. 2.2.1 in forming the $\pi$ model, Fig. 2.5. The Bode plot in Fig. 2.8 should also be compared with the corresponding plots of magnitude and phase for the same circuit obtained from the $C_t$ approximation, Fig. 1.7. The plots of magnitude and phase calculated from the six-by-six determinant are included in both figures to facilitate this intercomparison.

#### 2.3.3 Calculating Magnitude and Phase without Factoring

Often it is convenient to approximate the magnitude and phase of a function *without* first factoring the numerator and denominator polynomials. The basic idea of this approximation is that one term of a polynomial is often the dominant term over a given frequency range. For example, if

$$P(s) = a_0 + a_1 s + a_2 s^2 + a_3 s^3 \qquad (2.11)$$

then it is a fair approximation to assume:

$$\begin{aligned}
P(s) &= a_0, & 0 &< \omega < \frac{a_0}{a_1} \\
&= a_1 s, & \frac{a_0}{a_1} &< \omega < \frac{a_1}{a_2} \\
&= a_2 s^2, & \frac{a_1}{a_2} &< \omega < \frac{a_2}{a_3} \\
&= a_3 s^3, & \frac{a_2}{a_3} &< \omega < \infty
\end{aligned} \qquad (2.12)$$

This approximation is valid when all of the roots are on the real axis, as they are for the polynomials considered in this chapter, and will give reasonable answers even for poles somewhat off the axis, provided $(a_0/a_1) < (a_1/a_2) < (a_2/a_3)$. When both numerator and

## 50   Amplifier Calculations Using a π Model

denominator of the function are polynomials, we can sketch on logarithmic coordinates the magnitudes of the two polynomials separately, then subtract the two curves to determine the magnitude of the complete function.

As an example of this method, we use Eq. 2.12 to derive the approximate plot of $|A_v|$ versus $\omega$ from Eq. 2.7. The result is shown in Fig. 2.9. A comparison with the plot in Fig. 2.8a shows that the magnitude approximation is fairly good and is in error by a maximum (in this example) of about 20%.

The phase of $A_v$ can also be approximated from the unfactored polynomial. To accomplish this, we can take advantage of an im-

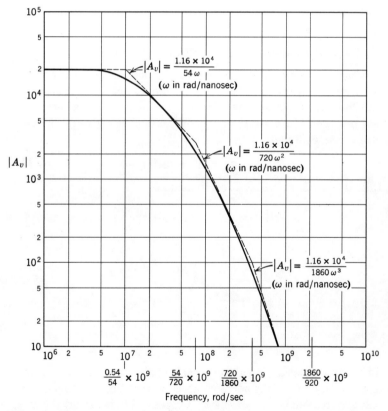

**Fig. 2.9.** Approximation to magnitude from nonfactored expression for $A_v$.

### Sec. 2.3 Performance Calculations Based on Transfer Function

portant property of polynomials in $j\omega$. We can write the denominator polynomial in Eq. 2.7 in the form

$$P(s) = (0.54 + 720s^2 + 920s^4) + (54s + 1860s^3)$$
(s in nanosec$^{-1}$)
(2.13)
$$P(j\omega) = (0.54 - 720\omega^2 + 920\omega^4) + j\omega(54 - 1860\omega^2)$$
($\omega$ in rad/nanosec)

Thus $P(j\omega)$ can have a phase of $0°$ or $\pm 180°$ only if

$$(54 - 1860\omega^2) = 0 \tag{2.14}$$

and can have a phase of $\pm 90°$ only if

$$0.54 - 720\omega^2 + 920\omega^4 = 0 \tag{2.15}$$

Solving these equations, we find

$$\angle A_v = \begin{cases} -90° \\ -270° \end{cases} \text{ for } \omega = 3.3 \times 10^7 \text{ or } 8.84 \times 10^8 \text{ rad/sec}$$
(2.16)
$$\angle A_v = -180° \quad \text{for } \omega = 1.71 \times 10^8 \text{ rad/sec}$$

These values check closely with the phase plot of Fig. 2.8b.

To illustrate a second method of finding $|A_v(j\omega)|$ from the unfactored polynomial, we first form $|A_v(j\omega)|^{-2}$ from Eq. 2.7

$$|A_v(j\omega)|^{-2} =$$

$$0.75 \times 10^{-8} [(0.54 - 720\omega^2 + 920\omega^4)^2 + \omega^2(54 - 1860\omega^2)^2]$$

$$= 0.22 \times 10^{-8} + 1.6 \times 10^{-5}\omega^2 + 2.4 \times 10^{-3}\omega^4$$

$$+ 1.6 \times 10^{-2}\omega^6 + 6.4 \times 10^{-3}\omega^8 \tag{2.17}$$
($\omega$ in units of rad/nanosec)

In Fig. 2.10, lines are drawn on a log-log plot corresponding to individual terms in Eq. 2.17, and then added algebraically to yield $|A_v(j\omega)|^{-2}$. Note the *inverted scale* for $|A_v(j\omega)|^{-2}$, which allows us to find $|A_v(j\omega)|$ without replotting. The figure also shows that only about three terms in Eq. 2.17 are significant over any limited frequency range. Thus, up to frequencies of the order of $\omega = 10^7$ rad/sec, we are probably justified in neglecting all but the first

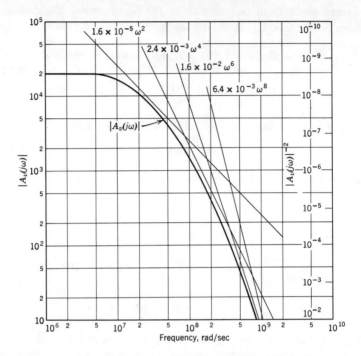

**Fig. 2.10.** Finding $|A_v(j\omega)|$ from $|A_v(j\omega)|^{-2}$

three terms. This is another way of justifying the approximation to the 0.707 frequency given by Eqs. 1.16 and 1.17.

## 2.4 SIMPLIFIED MODEL INCLUDING $R_e'$

As a second example of model simplifications appropriate for nodal analysis, we develop in this section a simple model for a transistor with an unbypassed emitter resistor $R_e'$, as shown in Fig. 2.11a. As we shall see, inclusion of $R_e'$ compensates for the attenuation due to the finite transverse dimension of the base, thereby improving transistor performance above the frequency $\omega_b$.

### 2.4.1 *Development of Model*

The first step in deriving a simplified model is to find the $Y$ matrix for the circuit of Fig. 2.11b, which contains only the ele-

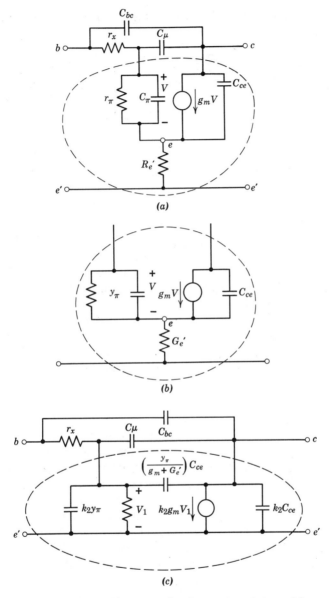

**Fig. 2.11.** Method of absorbing external emitter resistor into model.
$$k_2 = G_e'/(g_m + G_e') = 1/(1 + g_m R_e')$$

ments within the dashed contour in Fig. 2.11a. By direct calculation we find

$$[Y'] = \begin{bmatrix} \dfrac{y_\pi(G_e' + sC_{ce})}{g_m + G_e' + y_\pi + sC_{ce}} & \dfrac{-y_\pi sC_{ce}}{g_m + G_e' + y_\pi + sC_{ce}} \\ \dfrac{g_m G_e' - y_\pi sC_{ce}}{g_m + G_e' + y_\pi + sC_{ce}} & \dfrac{G_e' sC_{ce} + y_\pi sC_{ce}}{g_m + G_e' + y_\pi + sC_{ce}} \end{bmatrix} \quad (2.18)$$

Below $\omega_T/3$, we can neglect $y_\pi + sC_{ce}$ in comparison with $g_m + G_e'$ and introduce errors no larger than $-6\%$ in amplitude and $-19°$ in phase. On this basis, $[Y']$ can be approximated by the circuit within the dashed contour in Fig. 2.11c, where

$$k_2 = \frac{G_e'}{g_m + G_e'} = \frac{1}{1 + g_m R_e'} \quad (2.19)$$

Because $C_{ce}$ is generally quite small, it often turns out that the feedback admittance $y_\pi sC_{ce}/(g_m + G_e')$ in Eq. 2.18 and Fig. 2.11c can be neglected below $\omega_T/3$.

Thus we have developed a model which has the same *form* as a model containing only $r_\pi$, $C_\pi$, $C_\mu$, $C_{ce}$, and $g_m$, but which now *includes the effect of* $R_e'$ *for frequencies up to about* $\omega_T/3$. If desired, we can complete the $y$ formulation by applying the methods discussed in Sec. 2.1 to approximate the effect of $r_x$.

### 2.4.2 Example

As a numerical example of this method of incorporating $R_e'$ into the transistor model, let us take the transistor discussed in Sec. 2.1, and add a 70-ohm resistance to the emitter circuit. The complete hybrid-$\pi$ model with added $G_e'$ is shown in Fig. 2.12a. The simplified model is developed in the figure in three steps. First, the part within the dashed contour is modified by absorbing $G_e'$, in accordance with Fig. 2.11c. The result is shown in Fig. 2.12b. Note that we have neglected the small feedback element in drawing this figure. The $Y$ matrix for this portion of the circuit is:

$$[Y_1] = \begin{bmatrix} 0.11 + s\,8.4 & 0 \\ 10 & 0.3 \end{bmatrix}$$

## Sec. 2.4 Simplified Model Including $R_e'$    55

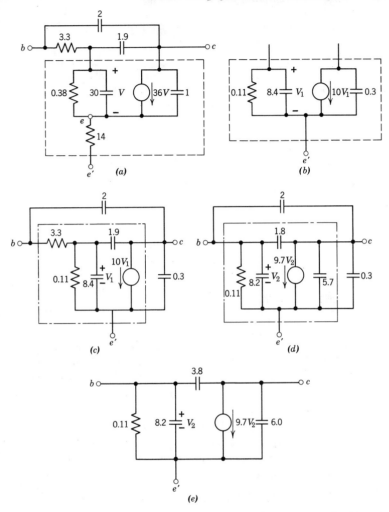

**Fig. 2.12.** Example of absorption of $R_e'$ and $r_x$ to form a $\pi$ model. Units of millimhos and picofarads.

The model with $R_e'$ absorbed in this manner is shown in Fig. 2.12c.

We are now in a position to absorb $r_x$, following the procedure of Sec. 2.1. To do this, we separate out that portion of the model in c within the dash-dot contour, and write the corresponding matrix on the basis of Eq. 2.4:

$$[Y_2] = \begin{bmatrix} k_1(0.11 + s\,10.3) & -k_1 s\, 1.9 \\ k_1(10 - s\ 1.9) & k_1 s\, 1.9 + \dfrac{10.1}{3.4} s\, 1.9 \end{bmatrix}$$

where $k_1 = \dfrac{g_x}{g_x + g_\pi} = \dfrac{3.3}{3.4}$.

Multiplying out, we obtain

$$[Y_2] = \begin{bmatrix} 0.11 + s10 & -s\,1.8 \\ 9.7 - s\ 1.8 & s\, 7.5 \end{bmatrix}$$

The resulting circuit is shown in Fig. 2.12d.

Now the extrinsic capacitors can be combined with the intrinsic to yield the model shown in Fig. 2.12e. Note that $R_e'$ has, in effect, increased the base cut-off frequency to:

$$\omega_b' = \frac{g_x + k_2 g_\pi}{k_2 C_\pi + k_2 C_\mu} = \frac{3.4}{10.3} \times 10^9 = 2.6 \times 10^8 \,\text{rad/sec}$$

That is, $f_b' = 48$ mc. Thus the final model shown in Fig. 2.12e provides a fairly good approximation to the transistor with a 70-ohm emitter resistance, an approximation which is valid up to about $f_b'/3 = 16$ mc. Recall from Sec. 2.1 that the corresponding $\pi$ model without $R_e'$ is good only to 6 mc.

The values in the Fig. 2.12e could have been approximated with very little effort. Note that absorbing an emitter resistance and base resistance does not appreciably affect either $g_m/y_\pi$ or the total collector-to-base capacitance. Also, $g_m$ (final) is approximately equal to $k_2 g_m$ (original), so the only parameter which is at all difficult to calculate is $C_{ce}$ (final).

## 2.5 SUMMARY

To summarize the model discussions in Chapters 1 and 2, we present in Table 2.2 a list of transistor models and the upper frequency limit of each model. The limit is chosen on the basis that the model can be expected to introduce errors of the order of 10% in amplitude and 20° in phase at this frequency.

## TABLE 2.2

| Model | Upper Frequency Limit |
|---|---|
| Hybrid-$\pi$ with extrinsic capacitance (Fig. 2.1) | $\omega_T = \dfrac{g_m}{C_\pi + C_\mu}$ |
| Intrinsic hybrid-$\pi$ (Fig. 1.2) | $\omega_b = \dfrac{g_x + g_\pi}{C_\pi + C_\mu}$ |
| $\pi$ model (Fig. 2.4) | $\omega_b/3$ |

Recall also that $C_t$ calculations of forward gain and input impedance (Sec. 1.5) are valid to about $3\omega_h$. [The frequency $\omega_h$, however, is not a property of the transistor alone, but also depends on the resistor values in the circuit. We shall see in Chapter 5 that in iterative stages it can have values from below $\omega_\beta(=\omega_T/\beta_0)$ to $\omega_b$.]

## PROBLEMS

**P2.1** A model for a two-stage amplifier, neglecting base resistance, is shown in Fig. 2.13. Element values are in millimhos and picofarads.
(a) Write the admittance matrix and evaluate the gain $A_v = A_0/Q(s)$.
(b) Determine the pole locations and sketch the asymptotic form of $A_v(j\omega)$.

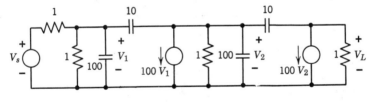

Fig. 2.13

**P2.2** Calculate the bandwidth $\omega_h$, and $A_v(j\omega)$ out to $\omega \cong 3\omega_h$, for the circuit of Fig. 2.13, using the methods of Chapter 1. Compare your answers with those of Problem P2.1.

**P2.3** A model of a two-stage amplifier is shown in Fig. 2.14. In this case both base resistance and overlap-diode capacitance have been explicitly included. Calculate and plot $A_v(j\omega)$. What is the range of validity of your solution? Element values are in millimhos and picofarads.

**P2.4** Calculate the bandwidth $\omega_h$, and $A_v(j\omega)$ out to $\omega \cong 3\omega_h$, for the circuit of Fig. 2.14, using the methods of Chapter 1. Compare the results with those of Problem P2.3.

## 58  Amplifier Calculations Using a π Model

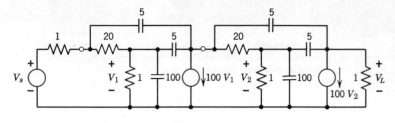

Fig. 2.14

**P2.5** For the circuit in Fig. 1.8, with values given in Problem P1.1, use the methods discussed in Chapter 2 to find a more accurate estimate of $A_v(j\omega)$, $\omega_h$, and the four lowest poles of the high-frequency group. Where appropriate, compare answers with those obtained in Problem P1.1.

**P2.6** Repeat Problem P2.5 for the values given in Problem P1.2.

**P2.7** The circuit of a two-stage transistor amplifier is shown in Fig. 2.15.
  (a) Estimate the quiescent operating points of the two transistors. Assume that both transistors have $\beta_F = 100$ and $I_{CO}$ is negligible. Make reasonable approximations.
  (b) Assume that the transistors can be represented by a hybrid-$\pi$ model which has the following element values at an operating point of $I_C = 1$ ma, $V_{CE} = 5$ volts: $r_x = 25$ ohms, $r_\pi = 2.5$ k, $r_o = 40$ k, $C_\pi = 20$ pf, $C_\mu = 2$ pf, $g_m = 0.04$ mho. Furthermore, assume that

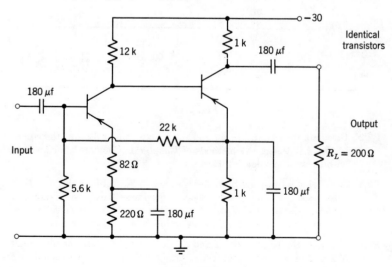

Fig. 2.15

the parameter values at the actual operating points of the two transistors can be determined by scaling $g_m$, $C_\pi$, $1/r_\pi$ as $|I_C|$, and leaving $r_x$ and $C_\mu$ unchanged. Show that the complete small-signal model of the amplifier in the high-frequency region (i.e., where voltage drops across bypass and coupling capacitors are negligible) is as shown in Fig. 2.16. Determine values for all of the circuit elements in this model.

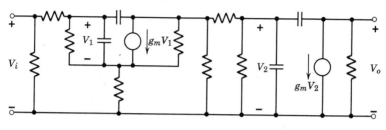

Fig. 2.16

(c) Convert to a $\pi$ representation.
(d) Compute the voltage gain $A_v = (V_o/V_i)$. Construct a pole-zero diagram for $A_v$ and sketch $|A_v|$ and $\angle A_v$ versus log $\omega$ in the frequency range of interest. Determine the mid-band gain and the bandwidth of this amplifier.
(e) It is desired to reduce the mid-band voltage gain of this amplifier to 20. This can be done by shunt loading between stages as shown in Fig. 2.17. Determine the proper values for $R_S$ to achieve the desired over-all gain. Compute the voltage transfer function and sketch both the pole-zero diagram and the gain and phase versus frequency.

**P2.8** Figure 2.18a shows a two-stage amplifier designed to give an 80-volt output swing for driving the grid of a cathode-ray tube. The high gain

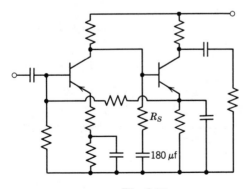

Fig. 2.17

of the output stage makes $C_\mu$ of the output transistor very important, so a 50 k resistor from collector to base is used to minimize the effect of this capacitance. An assumed high frequency incremental model for the circuit is shown in Fig. 2.18b. Element values are in millimhos and 100 picofarads.

(a) Write the exact admittance matrix corresponding to this model using $V_1$, $V_2$, and $V_L$ as the independent variables.

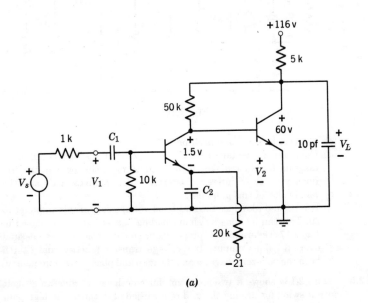

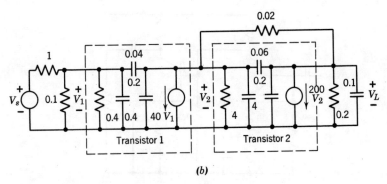

Fig. 2.18

(b) Indicate any approximations that are appropriate for calculating the voltage gain $A = V_L/V_s$ up to about $\omega = 10\omega_h$.
(c) Evaluate the expression for $A$ and compute approximate pole locations.
(d) Compute $\omega_h$ and determine which capacitor has the greatest effect on this value by calculating the open-circuit time constants.

# 3

# Feedback Amplifier Concepts

## 3.0 INTRODUCTION

The idea of feedback will not be new to the reader. The feedback situation occurs constantly about us. So far in this volume we have encountered it in several places without using the concept explicitly: e.g., the feedback effects of the $C_\mu$ of the transistor. So, clearly, the feedback point of view is not a necessary one. We can always analyze an amplifier system by using conventional circuit analysis without introducing feedback concepts at all, and in many cases this technique leads to adequate design procedures. In some situations, however, the feedback approach is very convenient, because it simplifies the problem of visualizing how certain modifications will affect amplifier performance.

The feedback method is used in this chapter principally in relation to two problems associated with cascaded amplifiers: the variation of gain level with changes of transistor- and circuit-parameter values; and the nonlinear relation between output and input for large output signals. Problems of frequency response and stability will be taken up in the next chapter.

The feedback method can be visualized for the present as follows. We start with an amplifier and feed back to the input a signal precisely related to the output. With this fed-back signal, we are able to modify the over-all performance of the system. We emphasize this notion with the general block diagram shown in Fig. 3.1. In the figure, a basic high-gain amplifier, a precision attenuator, and a comparator are shown. The total or net input signal to the basic amplifier consists of the actual input signal combined with a precisely attenuated version of the output signal. The combination is performed by the comparator, the output of which we may think of as an "error signal."

Presumably, what we want from the whole amplifier system is an output which is exactly a scaled-up version of the input, with an accurately known scale factor and no distortion of signal shape. As we suggested before, the basic amplifier suffers variations in gain for various reasons, and introduces signal distortion at the desired output levels. Its only compensating advantage is that it has much more gain than the ultimate scale factor we want between the input and desired output. We are willing to trade the extra gain for more accuracy. The feedback arrangement of Fig. 3.1, applied in such a way as to create *negative* or *inverse* feedback, can do this, under appropriate circumstances.

The simplest negative or inverse feedback situation is the one that does not involve time delays in any of the blocks. Let the precision attenuator have an attenuation ratio denoted by $f$, where $f$ is less than one, as shown in Fig. 3.1. Hence, at each instant the comparator compares the actual input time function $I(t)$ with the output multiplied by $f$. If the feedback is negative, this comparison is made by subtraction, and the difference becomes the error signal that drives the basic amplifier; thus

$$E(t) = I(t) - fO(t)$$

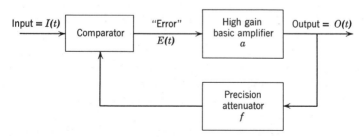

Fig. 3.1. General block diagram of a feedback amplifier.

Because the error signal drives the basic amplifier, however,

$$O = aE$$

Accordingly, as the gain $a$ of the basic amplifier is made increasingly large, it takes an ever smaller value of $E$ to produce a given size output $O$. Thus the attenuated output $fO$ fed back becomes ever closer to the value of the input $I$ at each moment, and the difference can be made arbitrarily small for large enough basic amplifier gain. Evidently, then,

$$O \underset{a \to \infty}{\to} \frac{I}{f}$$

which means that the gain of the over-all system approaches the reciprocal of the attenuation of the precision feedback attenuator, and is virtually *independent* of the original gain $a$. Under these circumstances, we have achieved a desensitization of the over-all amplifier gain function to changes in $a$.

Moreover, because $E = O/a$ and, therefore,

$$fO + \frac{O}{a} = I$$

it is clear that the basic amplifier gain $a$ must be large compared to $1/f$ in order to achieve the advantages of *desensitization of the amplifier gain function*. In other words, these advantages accrue only when the gain *sacrificed* by the feedback scheme is *very large*, that is, $(1/f) \ll a$, and success depends critically upon the precision attenuator. For proper comparison, that is for proper modification of the net input signal to the basic amplifier, we need a precisely known version of the output signal. If the attenuator and comparator do not provide this, we obtain an improper error signal. We are fortunate in this regard because precise attenuators can be made with passive $R$, $L$, and $C$ components. As we have seen, the over-all gain of a strongly desensitized amplifier is nearly equal to the inverse of the loss function provided by the feedback attenuator.

Let us examine in more detail some of the foregoing ideas, in terms of specific electrical network arrangements.

## 3.1 ELEMENTARY PROPERTIES OF FEEDBACK AMPLIFIERS

General feedback principles can be applied to almost any circuit. But, for simplicity, we start by discussing a circuit that consists of a unilateral basic amplifier which does not load, and is not loaded by, the passive feedback network. Figure 3.2 shows a typical electrical block-diagram representation of one type of feedback amplifier which has these properties. Because eventually we wish to discuss the effect of feedback on nonlinearities, the present analysis must be carried out in the time domain. Hence we use *instantaneous incremental* variables $v_I$, $v_F$, etc. Accordingly we also assume that the system blocks are without delays or other frequency response limitations, and focus attention on the system properties which are not sensitive to this assumption.

First, let us take up the case in which this system is *all linear*, but in which the voltage gain, $a_v$, of the transistor amplifier is presumably controlled by undependable transistor parameters; hence, it is subject to considerable variation. We assume that $a_v$ is a much larger gain than we actually need. In Fig. 3.2, we have represented the required passive attenuator by a circuit with voltage "gain" $f_v(<1)$ as measured from right to left. The "impedance" conditions shown in the figure assure that $a_v$ and $f_v$ are independent of loading, and guarantee that no current flows at the $v_F$ terminals. Thus the open-circuit voltage gain $f_v$ describes completely the feedback network, and there is no need to consider transmission of signals from left to right through it.

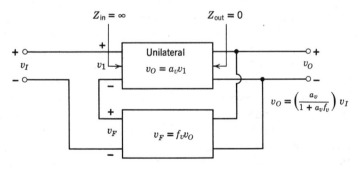

**Fig. 3.2.** Electrical block diagram of an idealized feedback amplifier.

The behavior of the system is then described simply by the relations

$$v_O = a_v v_1 = a_v(v_I - v_F) = a_v(v_I - f_v v_O)$$

from which we find, for the over-all voltage gain, the result

$$v_O = A_v v_I = \left(\frac{a_v}{1 + f_v a_v}\right) v_I \tag{3.1}$$

Evidently, $A_v$ approaches $1/f_v$ as $a_v$ approaches infinity, and this again illustrates our point that the resulting gain approaches the reciprocal of the attenuator "gain" when the basic amplifier gain becomes large.

In Eq. 3.1, the feedback is said to be *negative* or *inverse* if $a_v$ and $f_v$ are of the *same* algebraic sign, because then $v_I$ and $v_F$ oppose each other in producing $v_1$. Under these conditions, an increase in $a_v$ at fixed $v_I$ tends to increase $v_O$, increase $v_F$, and *decrease* $v_1$—which thus tends to decrease $v_O$, thereby opposing the effect of the initial change in amplifier gain. In general, the feedback is of *inverse* polarity if a small increase in the magnitude of $f_v$ from zero, in whatever polarity is being considered, *reduces* $|A_v|$ *below* $|a_v|$. Otherwise, the feedback is said to be *positive*.

To examine in detail the additional effects of feedback on such factors as system nonlinearity and other signal-distorting causes, we will use the dc amplifier given in Fig. 3.3a as the basis of discussion. This circuit is a practical (approximate) embodiment of the block diagram of Fig. 3.2. It consists of a two-stage, differential, direct-coupled amplifier with an $f_v$ which is independent of frequency and which can be controlled by a potentiometer $R_7$.

Here is a brief description of the operation of the amplifier in Fig. 3.3a. The constant-current source $I_O$ makes the sum of the emitter currents for the two transistors in the first stage practically independent of the collector-to-emitter voltages of these transistors. Thus the sum of the collector voltages for the first stage is stabilized with respect to the plus terminal of the power supply. This, in turn, stabilizes the sum of the emitter currents of the second stage; and hence, for zero input, if the circuit were perfectly symmetrical, all four transistors would have well-controlled quiescent currents. The net effect of using the current source $I_O$ (a practical form of which is shown as $T_3$ in the inset of Fig. 3.3a) is to

### Sec. 3.1 Elementary Properties of Feedback Amplifiers

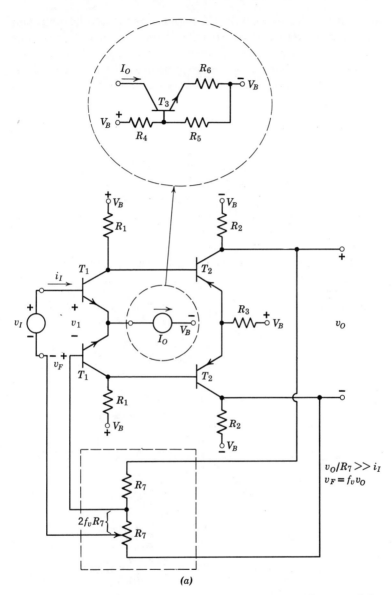

**Fig. 3.3** (a) Realization of Fig. 3.2 using a two-stage, direct-coupled, differential amplifier.

## 68  Feedback Amplifier Concepts

make it possible to analyze this amplifier as though the input pair of terminals were isolated from the output terminals in the same manner that a transformer primary is isolated from the secondary. As long as the input transistors are operating in their active regions, only the differential input voltage $v_1$ has a significant effect on the output.

### 3.1.1 *Effect of Feedback on Nonlinearities*

A circuit was constructed along the lines of Fig. 3.3a and the measured curve of $v_O$ versus $v_I$ was plotted for $f_v = 0$ and for $f_v = 0.05$. The results are shown in Fig. 3.3b. This figure shows three curves: $v_O$ versus $v_I$ for $f_v = 0$, $v_O$ versus $v_I$ for $f_v = 0.05$, and $v_O$ versus $v_I$ for an ideal amplifier with voltage gain 20 (which is the reciprocal of the precision attenuation 0.05).

Note that the curve for the over-all amplifier with $f_v = 0.05$ could also be found from a simple graphical construction by using the experimental $v_O$ versus $v_I$ plot for $f_v = 0$ and the straight-line

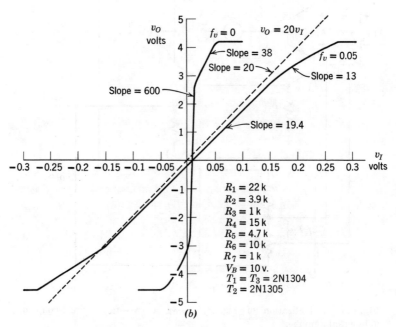

**Fig. 3.3.** (*Continued*) (b) Transfer curve for amplifier of Fig. 3.3a.

Sec. 3.1 Elementary Properties of Feedback Amplifiers    69

relation, $v_O = 20v_I$. To visualize this graphical construction of the $f_v = 0.05$ curve, it is helpful to focus attention on the input voltage required to achieve a given output voltage. Referring to Fig. 3.3a, we write:

$$v_I = v_1 + f_v v_O$$

We then identify $v_1$ as the input voltage which would be needed to produce $v_O$ with no feedback. That is, we put

$$v_1 = v_I|_{f_v = 0}$$

into the expression for $v_I$ to obtain

$$v_I = v_I|_{f_v = 0} + f_v v_O \qquad (3.2)$$

According to Eq. 3.2, we see that the curve for $f_v = 0.05$ in Fig. 3.3b can be derived as the horizontal addition of the curve for $f_v = 0$ and the straight line $v_I = 0.05\ v_O$. Thus *any departures of* $v_I|_{f_v = 0}$ *from a direct proportionality to* $v_O$ *are still present in* $v_I|_{f_v = 0.05}$ *to the same extent numerically*. But because the range of values of $v_I$ needed to obtain a given range of values of $v_O$ is larger with the feedback than without it, the linearity of the input-output relationship is better *on a percentage basis* when $f_v = 0.05$ than when $f_v = 0$. Whether or not this fact is in any sense helpful requires additional discussion.

### 3.1.2 *Reduction of Distortion*

Let us focus attention now on the effect of the feedback upon the distortion that arises from the nonlinearities at high output levels in the experimental curves of Fig. 3.3b. Note that the incremental gain of the basic amplifier changes from 600 to 38 when the magnitude of the output exceeds about 2.5 volts. This change in incremental gain occurs because one of the transistors in the output stage turns off. This action reduces the incremental voltage gain of the output stage to about 0.8 and increases the gain of the input stage to about 48, with a resulting over-all incremental gain $v_O/v_1$ of 38. Note that when feedback with $f_v = 0.05$ is applied to the amplifier, the $600:38 = 16:1$ change in gain (that occurred originally) is reduced to a $19.4:13 = 1.5:1$ change in gain. As we have pointed out before, the gain of the system with feedback is much less sensitive to the basic amplifier gain value than it would be without the feedback, *provided the basic gain remains large*. The over-all amplifier will therefore maintain reasonable linearity on a percentage basis, for output levels up to but not beyond the onset of saturation of the output transistor that remains "on."

But in view of the fact that this *percentage* improvement in the linearity has come about directly in proportion to the reduction in gain, there is considerable question whether any over-all improvement in performance has resulted; presumably the gain was the main reason for the making of the amplifier in the first place!

The point is, having applied the feedback, and having correspondingly reduced both the gain and the percentage distortion, we can now add new amplifier stages at the *input* to recover the lost gain. And these stages, *working at smaller signal levels*, will not necessarily reintroduce the same old distortion again. So we reduce the amplification function of the large-signal portion of the system in order to improve its large-signal fidelity, and then transfer the job of providing gain to the small-signal portions of the system, where the distortion is less. In this type of problem, then, inverse feedback does lead to improvement in over-all system linearity, without sacrifice of *system* gain.

Reference to Fig. 3.3b again, however, shows that ultimately, near 4 volts of output voltage, the incremental gain drops to practically zero. The circuit of Fig. 3.3a shows that this behavior results from saturation of whichever transistor is "on" in the second stage. Under this condition, with zero incremental gain left in the basic amplifier, the feedback effectiveness must fail and remain ineffective at increasingly large input levels.

An interesting experiment to illustrate this point is to apply a sine wave to the amplifier in Fig. 3.3a and then observe the voltage between the bases of transistors $T_1$. When one of the $T_2$ transistors turns off, the feedback circuit tries valiantly to maintain a sine wave at the output, and hence the voltage $v_1$ increases rapidly. The result is that when observing $v_1$ we see a greatly distorted wave (Fig. 3.4) even though the output is reasonably sinusoidal. We can determine this same distorted input waveform graphically in Fig. 3.3b by assuming that $v_O$ is approximately sinusoidal (even exactly sinusoidal will do), and noting what $v_1$ must be. To actually generate the resulting $v_1$ waveform in the circuit, from a perfectly sinusoidal $v_I$ and the feedback voltage $v_F$, it must be acknowledged

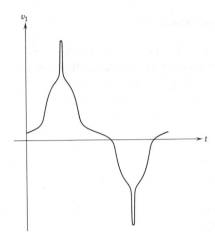

**Fig. 3.4.** Form of the *input* signal $v_1$ to the amplifier of Fig. 3.3a, when the *output* $v_O$ is sinusoidal, of peak voltage near 4.3 volts (see also Fig. 3.3b).

that $v_O$ is *not* exactly sinusoidal, but is somewhat flattened by the nonlinearities in the transfer curve. Figure 3.4 offers a very graphic proof *that, whereas feedback tends to produce over-all linearity between input and output, it tends to increase distortion for signals measured within the amplifier.*

This internal distortion can produce serious problems. Consider a circuit containing an emitter bypass capacitor. When feedback drives some transistors into saturation, the voltage across this capacitor may change significantly; and when the signal is removed, the operating points of the transistors will remain temporarily disturbed until the capacitor can discharge again. This internal overload in the presence of apparently linear over-all steady-state operation may mean that a short pulse of large amplitude can disable the amplifier. The amplifier may then be incapable of amplifying *small* pulses until the emitter bypass capacitor has returned to equilibrium conditions; and the time for this recovery may be very large. In Fig. 3.3a, a capacitor across $R_3$ can be used to give a startling demonstration of this large-signal blocking! As a general principle, then, we must *always* consider the effect of overload conditions when attempting to predict the behavior of a feedback amplifier.

### 3.1.3 *Extraneous Signals*

Another form of "nonlinearity" apparent in Fig. 3.3b occurs at a *low* (in fact, *zero*) output level, and is independent of signal size. Specifically, there is an "offset" such that the input is required to be +8 millivolts in order to make the output zero; and, as we pointed out in Sec. 3.1.1, effects like this 8-millivolt departure from linearity are numerically unchanged with and without the feedback. Of course, the change of input needed for an output of 3 volts is much greater with feedback (160 millivolts) than without it (about 8 millivolts). So the "offset" is about 50% of full-scale signal without feedback, and 5% with feedback. In this respect, the offset problem may seem similar to the distortion problem treated in the last section; but because it occurs at zero output level, and is *not a function of signal size*, it is in fact a *very different* problem.

First of all, the 8-millivolt offset arises from slight asymmetries in the amplifier (Fig. 3.3a) and, accordingly, would vary with such things as a change of transistors (of the same type), changes of supply voltages, aging of components, and variations of temperature. Thus different amplifiers built according to the same plan might well have different specific values of the offset voltage, but they would all be in the same general magnitude range. For this reason, an attempt to employ feedback to improve the situation leaves us with difficulty in recovering the gain. Any amplifier we build to put in front of the one with feedback, in order to bring up the system gain, can be expected to have about the same offset problem as the original amplifier, inasmuch as signal size has nothing to do with the effect. If the offset were any better, we should have used the preamplifier as the main one in the first place.

The fact that the preamplifier works at lower output levels than the main amplifier is no advantage here, where the trouble itself is independent of signal level. Use of inverse feedback for such cases is, therefore, not profitable.

Another way to analyze the action of feedback on amplifier defects of the offset type (i.e., those independent of signal amplitude) is to introduce an "extraneous signal" source in the amplifier at the place where the defect is observed. Figure 3.5a shows a revision of the block diagram of Fig. 3.2 to exhibit a signal source,

### Sec. 3.1 Elementary Properties of Feedback Amplifiers

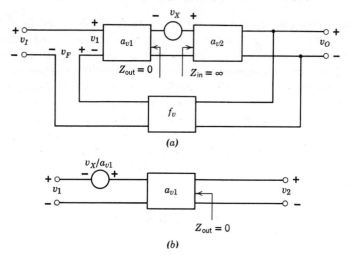

**Fig. 3.5.** (a) Feedback amplifier including extraneous signal source, $v_X$. (b) Equivalent input source that produces same output effects as $v_X$.

$v_X$, located somewhere between the input and output. To make the point clear, the situation is again idealized here by assuming that the output impedance of the first block is zero, and the input impedance to the second block is infinite. Under these conditions, a straightforward analysis yields this result:

$$v_O = \frac{a_v}{1 + a_v f_v}\left(v_I + \frac{v_X}{a_{v1}}\right)$$

where $a_v = a_{v1}a_{v2}$. But this implies that an "equivalent input source" $v_X/a_{v1}$ can be placed at the input to the first block (as shown in Fig. 3.5b), and reproduce at the output of the system the effect of $v_X$. The new source, often called "$v_X$ referred to the input," could as well have been obtained by using Thevenin's theorem.

On the basis of this analysis it would seem that we can, in principle, reduce the *effect* of $v_X$ simply by increasing $a_{v1}$. Unfortunately, as we pointed out, the amplifier that increases the gain $a_{v1}$ is expected to have extraneous signals of its own, such as noise and dc offset voltage, although it may not have a distortion defect attributable to large-signal operation.

If, on this basis, feedback will not help problems like the offset voltage, it might seem that we could simply introduce at the input a fixed voltage source of 8 millivolts to cancel the dc offset. In fact, any extraneous signal which is *predictable*, such as a fixed 60-cps component, could be cancelled by such a bucking signal. But we have already said enough to make it clear that the 8 millivolts is *not* predictable. Eventually, therefore, *we are always limited by our inability (or our unwillingness) to predict the nature of the extraneous signal*. Thus in Fig. 3.3a, if we really wish to reduce the unpredictable dc offset voltage (below the order of 8 millivolts suggested for it), we must resort not to feedback, nor to bucking signals, but to basic improvements, such as using matched transistors, precision resistors, and a controlled environment.

### 3.2 MORE DETAILED FORMULATION OF FEEDBACK VIEWPOINT

#### 3.2.0 *Introduction*

In our previous discussions connected with Figs. 3.2 and 3.5, we have limited our consideration to special cases in which the various networks involved do not load each other. This limitation has helped us to emphasize the main reasons for using the feedback point of view in linear and nonlinear systems, and to place in evidence the main advantages of this point of view. Now we must remove these loading restrictions, otherwise the feedback ideas we have discussed will hardly seem applicable to any real transistor circuits.

To proceed, let us focus upon the important aspects of feedback in *linear* systems. These aspects include reduction in system sensitivity to parameter variations, reduction in the effects of some extraneous signals, and improvement in fidelity or frequency response. The effects of feedback on nonlinearity are now not included, of course, *unless* the nonlinear phenomena can be regarded approximately as either changes in system parameters, or as generators of extraneous signals, in an otherwise linear system. Fortunately these approximate points of view are quite often useful.

For linear systems, it is convenient to think in terms of response at complex frequency $s$, and to describe the component networks in terms of impedance concepts. Of course, sinusoidal signals, represented by their complex amplitudes $V_o$, $V_i$, $I_i$, etc., are no

### Sec. 3.2 More Detailed Formulation of Feedback Viewpoint 75

longer in phase throughout the system, and accordingly time-domain waveforms change from point to point as a result of the frequency response of various portions of the system. We shall find eventually that the concepts of "positive" and "negative" feedback are not so clear anymore, and that the situation may change in this respect as a function of the frequency.

#### 3.2.1 *Some Typical Feedback Amplifier Configurations*

To apply the feedback point of view to a system, it is in most cases desirable to divide the system into two-port networks, one associated primarily with the basic high-gain amplifier (some of whose parameters are undependable), and the others associated primarily with the precision attenuator, or *feedback network*, whose parameters are dependable. If a clear enough division is not possible in a given case, there may well be a question about the practical utility of the feedback point of view for the problem at hand.

In Fig. 3.6 we show the four ways in which one basic amplifier two-port and *one* feedback network two-port can be joined together at their terminals.

The arrangement of Fig. 3.6c is detailed further in Fig. 3.7 to define notation, to demonstrate the method of handling the Norton equivalent representation of the source, and to demonstrate the inclusion of source and load conductances in with the basic amplifier. This rearrangement is helpful for the purpose of making an analysis which will place clearly in evidence the feedback point of view. In normal use, the output terminals at $V_o$ are open ($I_o = 0$). The feedback network may be thought of as "sampling" the output voltage $V_o$ across the load conductance $G_L$, and passing this information back as a *current* $I_{if}$, to be compared with $I_s$ in order to generate "error signal" current $I_{ia}$. Observe that the feedback is stopped if the load voltage $V_o$ is reduced to zero by shorting out $G_L$, but *not* if the load current $I_L$ is reduced to zero by opening up $G_L$. This is why we say that the feedback samples load voltage, rather than load current. Because current $I_{if}$ is the important response from the voltage excitation of the feedback network, we may refer to this configuration as *transadmittance feedback*.

Similarly, for comparison, Fig. 3.6a would be redrawn in the form of Fig. 3.8, where normally the output terminals at $V_o$ are

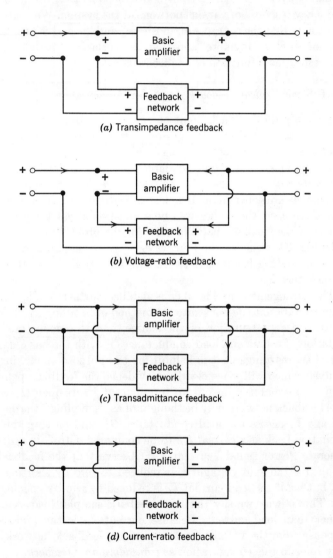

**Fig. 3.6.** Typical arrangements of two two-ports for a feedback amplifier. The polarities and arrows show the normal reference directions for terminal voltages and currents at the two-ports.

## Sec. 3.2 More Detailed Formulation of Feedback Viewpoint

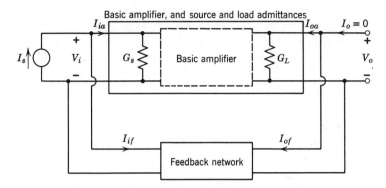

**Fig. 3.7.** A transadmittance feedback configuration, which samples load *voltage* $V_o$ and feeds back a related *current* $I_{if}$ against which to compare the source current $I_s$. Notice the treatment of the Norton equivalent source and the load conductance.

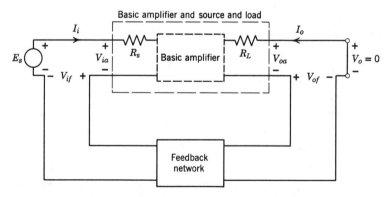

**Fig. 3.8.** A transimpedance feedback configuration, which samples load *current* $I_o$ and feeds back a related *voltage* $V_{if}$ against which to compare the source voltage $E_s$. Notice the treatment of the Thévenin equivalent source, and the load resistance, and compare with Fig. 3.7.

## 78  Feedback Amplifier Concepts

*shorted* ($V_o = 0$), as indicated by the solid line. Here the feedback is zero if the load current $I_o$ is made zero (by opening $R_L$), but not if $R_L$ is shorted. Hence, the feedback samples load *current* in this case, and delivers a *voltage* $V_{if}$ for comparison with $E_s$, to generate an "error signal" voltage $V_{ia}$. So this configuration may be called *transimpedance* feedback. For analogous reasons, Fig. 3.6b and d would be called *voltage-ratio* feedback and *current-ratio* feedback respectively.*

When joining networks as shown in Fig. 3.6, however, care must be taken to avoid connecting together leads which are forced by the individual networks to be at different potentials. Thus in Fig. 3.6a, a common ground in the basic amplifier could short out nodes on the top of the feedback network (unless its common ground were on top, of course); and in Fig. 3.6b, an amplifier common ground might well short out one terminal pair of the feedback network if the latter had a common ground. A balanced system, like Fig. 3.3a, may sometimes be used to avoid such difficulties if judicious joining of common grounds is not possible; or as a last resort, an isolation transformer may be used across one of the ports to block the undesirable circulating currents that otherwise tend to flow between the two networks.

We shall discuss primarily the various configurations indicated in Fig. 3.6, but it is well to point out that not only may we use combinations of these feedback arrangements, involving three two-ports (sometimes called "bridge feedback"), but we may also use combinations of $n$-port networks. However, even in such cases, the methods to be considered here will often be applicable with only slight modifications.

### 3.2.2 *Effect of Feedback on Impedance Levels*

Given that the feedback in each case is negative at the frequency of concern, the effects of different configurations on the input and output impedances of the system will be different. In general, whatever is "sampled" at the output by the feedback network tends to resist being changed.

---

* The four basic feedback configurations shown in Fig. 3.6 are sometimes referred to in the literature as series-series, shunt-series, shunt-shunt, and series-shunt, feedback.

## Sec. 3.2 More Detailed Formulation of Feedback Viewpoint 79

The sense in which this is so may be seen, for example, in Fig. 3.7, if we examine the output impedance by setting $I_s = 0$ and $G_L = 0$, and driving the output terminals with a voltage $V_o$. The "resistance to change" of $V_o$ shows itself in that the current $I_{oa}$, and therefore also $I_o$, is quite large. This effect results from "amplifier action" (i.e., by the dependent sources in the amplifier circuit model) actuated by the feedback excitation $I_{if}$. The extra current is in addition to that which would simply flow down through the amplifier output impedance if $V_o$ were applied in the absence of any signal fed back to the input. Therefore, the current $I_o$ is larger, and the output impedance is *lower* than it would be without the feedback.

The same argument applies to Fig. 3.6b, where again output *voltage* is sampled and, accordingly, the impedance measured at the output terminals shown is *reduced* by the feedback.

When the load *current* is sampled, as in Fig. 3.8, the application of a current $I_o$ instead of a short at the $V_o$ terminals (with $R_L = 0$ and the input source inactive) results in a large voltage $V_{oa}$ produced through amplifier action, and hence a large voltage $V_o$ at the output terminals. Thus the system output impedance is *high*. Similar reasoning applies to Fig. 3.6d.

At the input, the impedance can be calculated in a similar way. Consider Fig. 3.7 again, this time with $G_s = 0$. If the feedback is negative, the polarities must actually be such that $I_s$ is nearly equal to $I_{if}$, so that the "error" $I_{ia}$ is at all times kept small. Thus, application of a voltage $V_i$ results in a large current $I_{if}$ actuated through the feedback network from the amplifier *output*, in addition to that flowing down through the input impedance of the amplifier. Therefore, the input impedance is *small*. These comments also apply to Figs. 3.6c and d. Observe that this input impedance condition is consistent with the fact that the gain is reduced by the negative feedback. For a given current at the input terminals to the basic amplifier, the input voltage and output voltage are the same whether or not the feedback is operating (for the moment, we neglect loading effects of the feedback network). But with the negative feedback operative, the current at the *system* input terminals is *larger* than that of the basic amplifier, for the same voltage, which confirms that the input impedance is *reduced* by this type of feedback configuration.

The opposite effect occurs in case an input *voltage* is opposed by a feedback *voltage*, leaving only a small difference voltage to drive the high gain amplifier, as in Figs. 3.8, 3.6a, and 3.6b. The small difference voltage $V_{ia}$ in Fig. 3.8 (with $R_s = 0$) results in a small input current $I_i$, and a *high* input impedance presented to the system source ($E_s$ in Fig. 3.8).

A convenient summary of the impedance effects discussed above appears in Fig. 3.9, assuming a large amount of negative feedback and a high-gain basic amplifier for each case shown in Fig. 3.6. We include the fact that, for large feedback, the over-all gain is nearly equal to the inverse of the feedback attenuation, as pointed out previously and as illustrated by Eq. 3.1. Also in Fig. 3.9 is included the assumption that the reverse transmission of the system from output back to input is negligible compared to the forward gain. We shall have more to say about this matter shortly. The main point at the moment is that the figure makes it relatively easy to keep in mind the *essential* impedance and stabilization properties of the various elementary feedback configurations illustrated in Fig. 3.6.

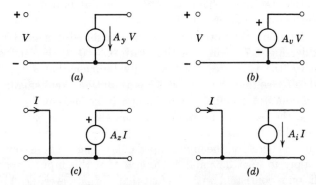

**Fig. 3.9.** Oversimplified summary of the limiting results of the inverse feedback configurations illustrated in Fig. 3.6. (*a*) Idealized result of large transimpedance feedback. The reliable $A_y^{-1} \cong$ open-circuit reverse transimpedance of feedback network. (*b*) Idealized result of large voltage-ratio feedback. The reliable $A_v^{-1} \cong$ open-circuit reverse voltage ratio of feedback network. (*c*) Idealized result of large transadmittance feedback. The reliable $A_z^{-1} \cong$ short-circuit reverse transadmittance of feedback network. (*d*) Idealized result of large current-ratio feedback. The reliable $A_i^{-1} \cong$ short-circuit reverse current ratio of feedback network.

### Sec. 3.2 More Detailed Formulation of Feedback Viewpoint

For mnemonic purposes in this connection, it may be helpful to recall that the over-all reliable gain with large feedback approaches the *reciprocal* of the feedback network attenuation. So trans*admittance* feedback will result in a stabilized trans*impedance* (Figs. 3.6c and 3.9c), trans*impedance* feedback will result in a stabilized trans*admittance* (Figs. 3.6a and 3.9a), and voltage-ratio or current-ratio feedback will result, respectively, in stabilized voltage gain and current gain. Moreover, the impedance levels are modified exactly as we would expect: a stabilized voltage gain must come with an infinite input impedance (certainly not with a short circuit!) and a zero output impedance (certainly not with an infinite one!), and similarly for the other cases. However, it is important to point out that *combination* feedback methods may produce very interesting and different results, such as precision finite and nonzero values of input or output impedance, along with reliable gain.

Of course, there are practical limits upon the gain of any amplifier, and therefore upon the amount of feedback that can be applied to it. So we must be able to deal quantitatively with the effects of only moderate amounts of feedback. This requires a somewhat more careful analysis of the feedback configurations than we have made thus far. The objective of this analysis will be to cast the result for the over-all gain in the "feedback form" of Eq. 3.1, in such a way that unreliable and reliable components in the system are clearly assignable to the appropriate separate factors. Only in this way can the feedback point of view help us visualize how to control explicitly the compromise between gain and desensitivity.

#### 3.2.3 Analysis in Feedback Form

Let us begin with Fig. 3.6c, redrawn in Fig. 3.10. This transadmittance feedback configuration is often especially convenient because it lends itself directly to the use of networks and amplifiers with common grounds. We have seen that, aside from the source and load, this transadmittance feedback configuration approaches a precision *transimpedance* (Fig. 3.9c) at large values of negative feedback, and this suggests that we discuss first the over-all open-circuit transimpedance of the system, $V_o/I_s$ for $I_o = 0$. If the actual source is not a current source in shunt with a conductance, we can always form the Norton equivalent.

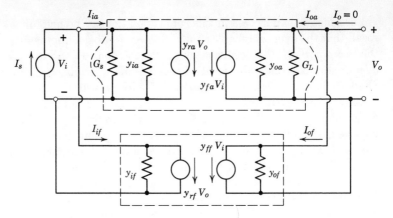

**Fig. 3.10.** Transadmittance feedback configuration.

If we are to be able to describe the whole system in Fig. 3.10 conveniently in terms of the parameters of each network alone, it would be helpful to choose as *independent* variables for *each* network the variables which are constrained to be equal when the networks are connected together in the feedback configuration involved. In this case, these variables are the input and output voltages, so the appropriate network parameters are the short-circuit admittances. For the basic amplifier we denote them by $y_{ia}$, $y_{fa}$, $y_{ra}$, $y_{oa}$, and for the feedback network by $y_{if}$, $y_{ff}$, $y_{rf}$, $y_{of}$. In accordance with standard two-port notation, the first subscript identifies *i*nput, *f*orward-transfer, *r*everse-transfer, and *o*utput functions. The second subscript identifies the *a*mplifier or *f*eedback network. Thus for the basic amplifier *and* source *and* load we have:

$$I_{ia} = (y_{ia} + G_s)V_i + y_{ra}V_o \tag{3.3a}$$

$$I_{oa} = y_{fa}V_i + (y_{oa} + G_L)V_o \tag{3.3b}$$

and for the feedback network:

$$I_{if} = y_{if}V_i + y_{rf}V_o \tag{3.4a}$$

$$I_{of} = y_{ff}V_i + y_{of}V_o \tag{3.4b}$$

Therefore, by addition of Eqs. 3.3 and 3.4 we find for the overall system:

$$I_s = I_{ia} + I_{if} = (y_{ia} + G_s + y_{if})V_i + (y_{ra} + y_{rf})V_o \tag{3.5a}$$

$$I_o = I_{oa} + I_{of} = (y_{fa} + y_{ff})V_i + (y_{oa} + G_L + y_{of})V_o \tag{3.5b}$$

## Sec. 3.2 More Detailed Formulation of Feedback Viewpoint

Condensing the notation to make $Y_i = y_{ia} + y_{if}$, etc., and noting the open-circuit condition $I_o = 0$ in Fig. 3.10 for normal operation as an over-all amplifier with load $G_L$, we find from Eq. 3.5b that

$$\frac{V_o}{V_i} = \frac{-Y_f}{Y_o + G_L} \tag{3.6}$$

This result employed in Eq. 3.5a leads to the over-all open-circuit transimpedance

$$\frac{V_o}{I_s} = \frac{-Y_f}{(Y_i + G_s)(Y_o + G_L) - Y_r Y_f}$$

There are a good many ways in which we might now try to put this result into a "feedback form." We shall select one that leads to convenient and useful results. Division in numerator and denominator by the first denominator term yields:

$$\frac{V_o}{I_s} = \frac{\left[\dfrac{-Y_f}{(Y_i + G_s)(Y_o + G_L)}\right]}{1 + Y_r \left[\dfrac{-Y_f}{(Y_i + G_s)(Y_o + G_L)}\right]} \tag{3.7}$$

Even Eq. 3.7, however, is open to a good many feedback "interpretations." For example, it can be interpreted as the form

$$\frac{V_o}{I_s} = \frac{a_z}{1 + f_y a_z} \tag{3.8}$$

in which *by definition* we take

$$f_y a_z \equiv -\left(\frac{Y_f}{Y_i + G_s}\right)\left(\frac{Y_r}{Y_o + G_L}\right) \tag{3.9a}$$

(The subscripts on $a$ and $f$ identify their electrical characters: $v$ for a voltage ratio, $y$ for a transfer admittance, $z$, for a transfer impedance, and $i$ for a current ratio.) The form in Eq. 3.9a happens to be the negative product of the short-circuit current gain of the *whole system* of Fig. 3.10, measured from left to right, by the same quantity measured from right to left; that is,

$$f_y a_z = -\left(\frac{I_o}{I_s}\right)_{V_o=0} \left(\frac{I_s}{I_o}\right)_{V_i=0}$$

But it can *also* be written:

$$f_y a_z \equiv -\left(\frac{Y_f}{Y_o + G_L}\right)\left(\frac{Y_r}{Y_i + G_s}\right)$$

$$= -\left(\frac{V_o}{V_i}\right)_{I_o=0}\left(\frac{V_i}{V_o}\right)_{I_s=0}$$

which is the negative of the product of the open-circuit voltage gains in both directions. Based upon either of these interpretations, $f_y a_z$ would be quite easy to measure directly.

Along with the foregoing identifications, it would seem reasonable to identify $f_y$ through the relation

$$f_y = Y_r \tag{3.9b}$$

which is clearly the short-circuit reverse transfer admittance of the whole system. Correspondingly, we might then identify $a_z$ as:

$$a_z = -\frac{Y_f}{(Y_i + G_s)(Y_o + G_L)} \tag{3.9c}$$

Equation 3.9c does not have a straightforward interpretation *in the general case*, but for the special case $y_{ra} = 0$, it is extremely useful, as we shall see below.

### 3.2.4 *Conditions Conducive to the Feedback Viewpoint*

There is one over-all circumstance under which the interpretations of all the feedback parameters in Eqs. 3.9 are both useful and at least tolerably convenient, namely, *in case it is possible to regard the high-gain basic amplifier as unilateral* ($y_{ra} \approx 0$). Actually there are two different effects in terms of which we must determine whether $y_{ra}$ is small enough to be negligible.

First, *the basic amplifier input admittance must be little affected by its load admittance.* To make the statement quantitative, we include in the loading on both ends of the amplifier the effects of the admittances $y_{if}$ and $y_{of}$ of the feedback network (Fig. 3.10) and we find for the amplifier under these load conditions:

$$Y_{in} = Y_i + G_s - \frac{y_{fa} y_{ra}}{Y_o + G_L}$$

## Sec. 3.2 More Detailed Formulation of Feedback Viewpoint

Conditions in the output will be unimportant only if the real and imaginary parts of the third term are negligible:

(a) $\left|\text{Re}\left[\dfrac{y_{fa}y_{ra}}{Y_o + G_L}\right]\right| \ll |\text{Re}[Y_i] + G_s|$

(b) $\left|\text{Im}\left[\dfrac{y_{fa}y_{ra}}{Y_o + G_L}\right]\right| \ll |\text{Im}[Y_i]|$

which in turn implies that:

$$|y_{fa}y_{ra}| \ll |Y_i + G_s|\,|Y_o + G_L|$$

Observe that the product $|y_{fa}y_{ra}|$ is usually smaller for a cascade of amplifier stages than for any one alone. So the above condition is likely to be valid for the cascade, even if it is doubtful for the individual stages.

But, in addition, we must also insist that *most of the signal carried back from system output to input travels through the feedback network, not through the basic amplifier.* This requires that if $y_{ra} = g_{ra} + jb_{ra}$ and $y_{rf} = g_{rf} + jb_{rf}$,

$$|g_{ra}| \ll |g_{rf}| \quad \text{and} \quad |b_{ra}| \ll |b_{rf}|$$

which implies that

$$|y_{ra}| \ll |y_{rf}|$$

Under these conditions, Eq. 3.9b becomes

$$f_y \approx y_{rf} \qquad (3.9d)$$

Notice that, because of Eq. 3.9d, the feedback factor (and, therefore, the ultimate gain with large feedback) depends *only* upon the reliable passive portion of the system. This is clearly an important feature of these results.

Finally, it is to be expected that the basic amplifier, rather than the passive feedback network, will supply most of the output signal. That is, the amplifier supplies the large gain. This makes it reasonable to assume that if $y_{fa} = g_{fa} + jb_{fa}$ and $y_{ff} = g_{ff} + jb_{ff}$,

$$|g_{fa}| \gg |g_{ff}| \quad \text{and} \quad |b_{fa}| \gg |b_{ff}|$$

which implies

$$|y_{fa}| \gg |y_{ff}|$$

and Eq. 3.9c becomes

$$a_z \approx \frac{-y_{fa}}{(Y_i + G_s)(Y_o + G_L)} \qquad (3.9e)$$

The interpretation of Eq. 3.9e, illustrated in Fig. 3.11a, is also important. We see that $a_z$ is the transimpedance of the *system* with $y_{ra}$, $y_{ff}$, and $y_{rf}$ set to zero. That is, it is the transimpedance of the basic amplifier assumed unilateral ($y_{ra} \approx 0$), working between useful source and load conductances $G_s$ and $G_L$; *but it is also shunted by the short-circuit terminal admittances of the feedback network* ($y_{if}$ and $y_{of}$). Whereas these are often negligible compared to $y_{ia} + G_s$ and $y_{oa} + G_L$, it is by no means always possible to guarantee such a condition in transistor circuits. So, whenever these loading effects do arise, we must understand how to handle them in the manner just illustrated.

In summary, we have been able to cast the rather general two-port transadmittance feedback situation of Fig. 3.10 into a simple and meaningful feedback form defined by Fig. 3.11a and by Eqs. 3.8, 3.9d, and 3.9e, which for convenience we repeat in slightly modified form:

$$\frac{V_o}{I_s} = A_z = \frac{a_z}{1 + f_y a_z} \qquad (3.10a)$$

$$a_z = \frac{V_L'}{I_s} = \left.\frac{V_o}{I_s}\right|_{\substack{y_{ra}=0 \\ y_{rf}=0 \\ y_{ff}=0}} = \frac{-y_{fa}}{(y_{ia} + y_{if} + G_s)(y_{oa} + y_{of} + G_L)} \qquad (3.10b)$$

$$f_y = \left.\frac{I_{if}}{V_o}\right|_{V_i=0} = y_{rf} \qquad (3.10c)$$

This form is predicated on two conditions: $y_{ra} \approx 0$, and $|y_{fa}| \gg |y_{ff}|$. Note that by properly accounting for the loading from $y_{if}$ and $y_{of}$, as in Fig. 3.11a, $a_z = V_L'/I_s$ *can be found directly from the circuit, without ever finding* $y_{fa}$, $y_{ia}$, *and* $y_{oa}$.

### 3.2.5 Desensitivity

With Eq. 3.10 at hand, we can substantiate more completely the desensitivity aspects of the negative feedback amplifier. The basic amplifier (Fig. 3.10) is considered to provide a net 180° phase

## Sec. 3.2 More Detailed Formulation of Feedback Viewpoint

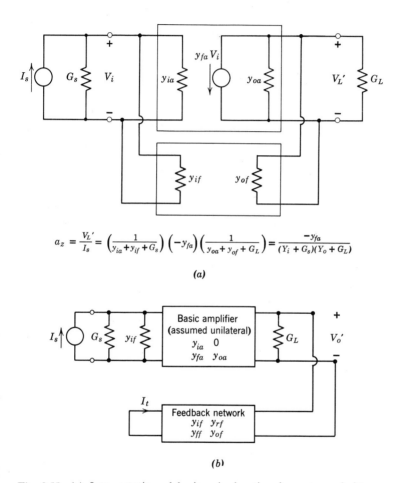

**Fig. 3.11.** (a) Interpretation of basic gain function for a transadmittance feedback amplifier (see also Fig. 3.7). (b) Interpretation of the loop transmission for a transadmittance feedback amplifier (see also Figs. 3.10 and 3.7). Note that on the basis of this figure, $a_z = V_L'/I_s$ or $T = I_t/I_s$ can be found directly from the circuit (often by inspection) without ever calculating $y_{fa}$, $y_{ia}$, and $y_{oa}$.

shift at midband ($y_{fa} = g_{fa} > 0$) (e.g., a three-stage common-emitter cascade), and we express $a_z$ as $-|a_z|$. If $f_y$ is purely resistive, as is often the case, $y_{rf} = g_{rf} < 0$. Thus, $f_y$ and $a_z$ are of the *same* sign, and the feedback current is out of phase with the input current. Notice that if the basic amplifier and/or the feedback network is adjusted to provide $a_z f_y \gg 1$, the over-all gain function is:

$$A_z \bigg|_{a_z f_y \gg 1} \cong \frac{1}{f_y}$$

For this situation, the gain function of the complete amplifier is approximately the reciprocal of the transmission function of the feedback network. The feedback network is made from passive elements which can be chosen to be very stable and precise, so its transmission function can be constant and precisely known. Clearly, then, we have desensitized the amplifier.

An example illustrates how good an approximation the foregoing provides. Let the low-frequency gain of the basic amplifier be $a_z = -100{,}000$ ohms. If the feedback function $f_y$ is adjusted for $-1/100$ mho, the product $a_z f_y = +1000$. From Eq. 3.10a, $A_z = -(100{,}000/1001)$ ohms, and $1/f_y = -100$ ohms. The over-all gain is equal to $1/f_y$ to within a fraction $1/1000$ (i.e., $1/a_z f_y$).

The desensitivity aspects of negative feedback actually pertain to the *percentage changes* in the gain of the basic amplifier, not the absolute level of gain. To bring this out more strongly, we take the differential of Eq. 3.10a, assuming that $f_y$ is constant.

$$dA_z = -\frac{da_z}{(1 + f_y a_z)^2}$$

Using Eq. 3.10a, we obtain

$$\frac{dA_z}{A_z} = \frac{1}{(1 + a_z f_y)} \left(\frac{da_z}{a_z}\right) \tag{3.11}$$

This very important result shows that a fractional change in $a_z$ leads to a fractional change of the over-all gain reduced by $1/(1 + a_z f_y)$. For the numbers above, if $a_z$ changes by 10%, the change in $A_z$ is only $(10/1001)\%$, or about $0.01\%$.

## Sec. 3.2 More Detailed Formulation of Feedback Viewpoint

### 3.2.6 Loop Transmission and Open Loop Gain

It is clear from the last section that the important quantity in the amplifier with negative feedback is $1 + a_z f_y$, or if we are to achieve significant densensitivity, simply $a_z f_y$. Algebraically, this quantity is defined simply by the product of Eqs. 3.10b and 3.10c, or by Eq. 3.9a with $y_{ra}$ and $y_{ff} = 0$:

$$a_z f_y = \left[ \frac{-y_{fa}}{(y_{ia} + y_{if} + G_s)(y_{oa} + y_{of} + G_L)} \right] y_{rf}$$

But Fig. 3.11b shows that this function can now be interpreted in a new way. It is the "current gain $I_t/I_s$ around the whole feedback loop" when the loop is "opened". This is accomplished by disconnecting the feedback network from the basic amplifier input, replacing it at the amplifier terminals by its short-circuit admittance $y_{if}$, and short-circuiting the now free terminals of the feedback network, where current $I_t$ will flow in response to $I_s$. With reference to Figs. 3.11a and b, this current gain may be written:

$$\frac{I_t}{I_s} = \left( \frac{V_o'}{I_s} \right) \left( \frac{I_t}{V_o'} \right) = \left( \frac{V_o'}{I_s} \right) y_{rf} = a_z f_y \equiv T \quad (3.12)$$

in which $T$ is called the *loop transmission*. As we have pointed out before, when the phase of the loop transmission $T$ is zero, the gain is *reduced* by the feedback and we have the ideal negative feedback situation.

On account of the method of "opening the loop" indicated in Fig. 3.11b, it is clear by comparing that figure with Fig. 3.11a that $a_z$ may also be called the *open-loop gain*, because the basic amplifier is loaded in the same manner in both figures. Thus we often say that the *closed-loop gain* $A_z$ is equal to the *open-loop gain* $a_z$, divided by one plus the *loop transmission* $T$.

$$A_z = \frac{a_z}{1 + T} \quad (3.13)$$

### 3.2.7 Input and Output Admittances

If in Fig. 3.10 we apply a voltage source $V_i = 1$ volt to measure the input admittance, we have $I_{ia} = G_s + y_{ia}$ because the basic amplifier is assumed to be unilateral. Also, the current $I_{if}$ is given by:

$$I_{if} = y_{if} + y_{rf} V_o \quad (3.14)$$

But because we neglect $y_{ff}$, we have $I_{of} = y_{of}V_o$, and

$$V_o = \frac{-y_{fa}}{G_L + y_{oa} + y_{of}} \quad (3.15)$$

Accordingly, from Eqs. 3.14 and 3.15, we find

$$I_{if} = y_{if} - \frac{y_{rf}y_{fa}}{G_L + y_{oa} + y_{of}}$$

Therefore, the total input current $I_\text{in} = Y_\text{in}$ is:

$$Y_\text{in} = I_{ia} + I_{if} = G_s + y_{ia} + y_{if} - \frac{y_{rf}y_{fa}}{G_L + y_{oa} + y_{of}}$$
$$= (G_s + y_{if} + y_{ia})(1 + T) \quad (3.16)$$

*The input admittance is increased from the value it would have with "no feedback" (in the sense $y_{rf} = 0$) by the same factor $1 + T$ that reduces the gain.*

Of course, normally the input admittance of the amplifier system would be defined as what the source sees, and $G_s$ is part of the source. So we should set $G_s = 0$ in Fig. 3.10 and Eq. 3.16 to get the proper value of $Y_\text{in}$ for the amplifier system alone. *But note that by this method $G_s$ appears in $T$ as well as in the first factor!* So the easiest way to find the input admittance of the amplifier *alone* may well be simply to subtract $G_s$ from the value given by Eq. 3.16. Either method gives the same result, of course.

As regards the output admittance, we think of applying $V_o = 1$ volt, with $I_s = 0$, and determining $I_o = Y_\text{out}$. Again, because we neglect $y_{ff}$, we have:

$$I_{of} = y_{of}$$

and because the basic amplifier is assumed to be unilateral, we find:

$$V_i = \frac{-y_{rf}}{G_s + y_{ia} + y_{if}}$$

It follows that:

$$I_{oa} = (y_{oa} + G_L) + y_{fa}V_i = y_{oa} + G_L - \frac{y_{rf}y_{fa}}{G_s + y_{ia} + y_{if}}$$

and that:

$$I_o = I_{oa} + I_{of} = Y_\text{out} = (G_L + y_{oa} + y_{of})(1 + T) \quad (3.17)$$

So the output admittance is also increased from its value "without feedback" (in the sense $y_{rf} = 0$) by the same factor $1 + T$ that reduces the gain. Again, the output admittance is normally defined without including $G_L$. So it should either be placed equal to zero in Eq. 3.17, and again this must be done in $T$ as well as in the first factor of Eq. 3.17, or $G_L$ should simply be subtracted from the result given by Eq. 3.17.

Both the input and output admittances have, in addition to their normal values, these values multiplied by the loop transmission. This circumstance stems from the simple fact that a source applied at either end of the system produces its normal response, plus that of a signal which proceeds completely around the loop and back to its origin. The idea of such a signal traversing the loop is really clearcut only if at least one element in the loop is unilateral.

## 3.3 EXAMPLES OF ELEMENTARY FEEDBACK ANALYSIS

### 3.3.0 *Introduction*

The general discussions we have made of feedback configurations and the feedback point of view are not always clear until they are applied to specific circuit arrangements. This we must now undertake to do.

### 3.3.1 *Example of Transadmittance Feedback*

In this section, a simple cascaded amplifier and a simple transadmittance feedback network are used to illustrate the points at hand. The system is shown in Fig. 3.12. For the feedback network, we use only the single resistor, $R_f$, which is drawn over a ground line to emphasize that we have the same two-port combination as in Fig. 3.10. For the transistors, we use the following parameters:

$$\beta_0 = 60, \qquad r_x = 50 \text{ ohms}$$

at $I_C = 10$ ma, $V_{CE} = 5$ volts.

We first calculate the value of the low-frequency gain of the basic amplifier. If we neglect $r_\mu$, $r_o$, and the loading of the feedback network, the current gain from $I_s$ to $I_1$ is:

$$-(60)\frac{R_s}{R_s + r_x + r_\pi} = -(60)\frac{50}{250} = -12$$

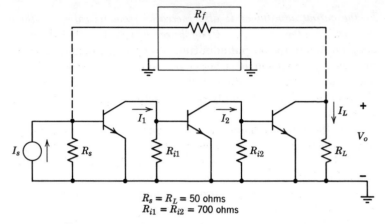

**Fig. 3.12.** Circuit of amplifier with transadmittance feedback. Biasing networks and power supplies have been omitted.

Because $R_{i1} = R_{i2}$, the current gains of the next two stages are the same and equal to $-(60)(700/900) = -46.7$. Thus the low-frequency current gain of the basic amplifier is $-(12)(46.7)^2 = -26,000$, and $a_z = -50(26,000) = -13 \times 10^5$ ohms.

Let us assume that we need a transimpedance $A_z$ of only 50,000 ohms. The extra gain can be used to achieve desensitivity. For the feedback network, $y_{rf} = -1/R_f$, and from Eq. 3.10c this is the feedback function.

$$f_y = -\frac{1}{R_f} \tag{3.18}$$

The loop transmission is:

$$T = a_z f_y = \frac{1,300 \times 10^3}{R_f}$$

By using the fact that we want $A_z$ to be $-50,000$ ohms, we can solve for $R_f$:

$$A_z = \frac{a_z}{1 + T}$$

$$T = \frac{a_z}{A_z} - 1 = \frac{13 \times 10^5}{5 \times 10^4} - 1 = 25$$

$$\therefore R_f = \frac{1,300 \times 10^3}{25} = 52 \text{ k}$$

### Sec. 3.3 Examples of Elementary Feedback Analysis 93

The input impedance to the system without feedback is just $R_s \parallel (r_x + r_\pi) = 50(200)/250 = 40$ ohms ($R_f$ is so large that it does not, in this case, load the input). So with the feedback, the impedance is lowered to:

$$R_{\text{in}} = \frac{40}{1+T} = \frac{40}{26} = 1.54 \text{ ohms}$$

The shunting effect of $R_s = 50$ ohms is very small:

$$G_{\text{in}} - G_s = (1.54)^{-1} - 0.02 = 0.650 - 0.02 = 0.63$$

or

$$R_{\text{in}}|_{R_s = \infty} = (0.63)^{-1} = 1.59 \text{ ohms}$$

The alternative calculation is more complicated, but let us see how it goes anyway.

The input impedance to the system with $R_s = \infty$ would have been $r_x + r_\pi = 200$ ohms without the feedback ($R_f$ is so large that it still does not load the input); but with feedback, a given current $I_L$ requires the same input voltage, but $(1 + T_{R_s = \infty})$ times as much input current (again $R_f$ does not appreciably load the input terminal). With $R_s = \infty$, $T$ is five times greater than before. This occurs because the first stage current gain is $-60$ instead of $-12$. So,

$$T_{R_s = \infty} = 5(25) = 125$$

Therefore, the input impedance is $200/(1 + T_{R_s = \infty}) = 200/126 \approx 1.59$ ohms, a value which agrees with the previous one.

As regards the output impedance, in this case $R_L$ supplies most of the loading (we have neglected $r_o$ and $r_\mu$ of the transistors). Thus the output impedance "without feedback" is just $R_L \parallel R_f \approx R_L = 50$ ohms. So with feedback, $T = 25$, and

$$R_{\text{out}} = \frac{R_L}{26} = 1.9 \text{ ohms}$$

The alternative calculation, again more complicated, has some worthwhile lessons which we shall not pursue here. It involves finding the output impedance when $G_L = 0$. (See Problem P3.7.)

At this point, we should go back and check our assumptions. First, we check whether at midband frequencies the input impedance of the whole basic amplifier is substantially independent of the load, in the sense that:

$$|y_{fa}y_{ra}| \ll |Y_i + G_s| \; |Y_o + G_L|$$

To determine approximately the transfer admittances, we observe that for each stage, $g_\mu$ (which we neglected) is of the order of $10^{-6}$ mho, and $g_m$ is of the order of 1 mho, whereas the interstage source and load impedance levels are approximately $R = 100$ ohms. The overall forward transfer admittance $y_{fa}$ is proportional to $g_m{}^3R^2$, whereas the overall reverse transfer admittance $y_{ra}$ is proportional to $g_\mu{}^3R^2$. Thus the product $|y_{fa}y_{ra}|$ is of order $10^{-10}$, which, when compared to the square of the general loading admittance level $(10^{-2})^2 = 10^{-4}$, is clearly negligibly small.

Next, the reverse transfer admittance of the feedback network dominates that of the basic amplifier because $r_\mu$ is of the order of megohms. The feedback through these resistors should be much less than the feedback through $R_f (= 52 \text{ k})$. Finally, we have a basic amplifier forward transadmittance without feedback of 26,000/40 (the current gain over the input impedance), or 650 mhos, which is certainly large compared to $y_{ff} = 1/R_f = 1/52{,}000 \approx 2 \times 10^{-5}$ mho.

Our assumptions are all well satisfied; therefore, we predict that placing $R_f$ across the amplifier will reduce the transimpedance level to the desired 50,000 ohms and we should have achieved a desensitivity to gain-level change of $1 + T = 26$. These results may not be obtained, however, with the actual amplifier. When the feedback resistor of 52 k is placed across the amplifier, the amplifier may oscillate. We have not discussed the phase shift of $a_z$ and, therefore, of $T$, at the higher frequencies. At these frequencies the negative feedback will change to positive feedback. This fact will be explored further in the next chapter. For the present we continue with examples of different feedback configurations.

### 3.3.2 Example of Transimpedance Feedback

The arrangement of Fig. 3.13a is sometimes called "series feedback," and is a special case of the two-port transimpedance feedback configuration of Fig. 3.8. Referring to Fig. 3.8, the proper *independent* variables (those that are the *same* for both networks when they are interconnected) are seen to be the *currents* $I_i$ and $I_o$.

### Sec. 3.3 Examples of Elementary Feedback Analysis

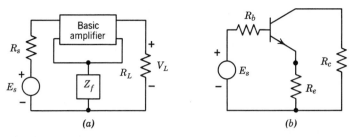

**Fig. 3.13.** Examples of transimpedance feedback configurations. (a) Single impedance "series feedback" network. (b) Emitter feedback network (bias supplies not shown).

The voltages are then the *dependent* variables, so the proper descriptions for the networks are the open-circuit self and transfer impedances of the amplifier and feedback networks, $(z_{ia}, z_{fa}, z_{ra}, z_{oa})$ and $(z_{if}, z_{ff}, z_{rf}, z_{of})$, respectively. Figure 3.9a shows that the stabilized quantity is the system *transadmittance* $A_y = I_o/E_s$. The whole problem is clearly dual to the one discussed in Eqs. 3.3 to 3.10, and the results must be dual in nature. In particular, we must find:

$$A_y = \frac{I_o}{E_s} = \frac{a_y}{1 + a_y f_z} = \frac{a_y}{1 + T} \quad (3.19a)$$

$$a_y = \frac{I_o}{E_s}\bigg|_{z_{ra}, z_{rf}, z_{ff}=0} = \frac{-z_{fa}}{(z_{ia} + z_{if} + R_s)(z_{oa} + z_{of} + R_L)} \quad (3.19b)$$

$$f_z = \frac{V_{if}}{I_o}\bigg|_{I_i=0} = z_{rf}$$

$$T = a_y f_z$$

For the simple case shown in Fig. 3.13a, we note that $z_{if} = z_{of} = z_{rf} = Z_f$, and the familiar common-emitter feedback circuit shown in Fig. 3.13b has $R_e$ playing the role of $Z_f$. The transistor, together with $R_s = R_b$ and $R_L = R_c$, is the basic amplifier. Let us estimate the desensitivity $1 + T$ for this arrangement.

For a numerical example, we take the following values: $R_c = 1$ k, $R_b = 1.7$ k, $R_e = 500$ ohms, $1/g_m = 2.5$ ohms, $\beta_0 = 50$, and $r_x = 50$ ohms. The low-frequency transadmittance of the basic amplifier is:

$$a_y = \frac{\beta_0}{R_b + R_e + r_\pi + r_x}$$

This checks with Eq. 3.19b if we realize that $z_{oa}$ is much greater than both $z_{of}$ and $R_L$, and that $z_{fa} = -\beta_0 z_{oa}$. For the values above, $a_y = 21 \times 10^{-3}$ mho. Then, inasmuch as $R_e = 500$ ohms, we have:

$$f_z = R_e = 500 \text{ ohms}$$

Therefore, $T = a_y f_z = 10.5$, and our desensitivity factor is $1 + T = 11.5$.

We can use the value of desensitivity to estimate the changes which occur if a parameter of the basic amplifier is changed. For example, let us assume that because of a temperature increase, the $\beta_0$ of the transistor increases by 10%. Accordingly $a_y$ changes by the same amount. Because of the desensitivity of 11.5, we expect the over-all gain $A_y$ to change by only $(10/11.5)\%$, or less than 1%.

Often the percentage change of $\beta_0$ is not small. To handle this case, we must derive a desensitivity expression which applies for large changes. Label the initial values of $a_y$ and $A_y$, $a_{y1}$ and $A_{y1}$, respectively, and the final values $a_{y2}$ and $A_{y2}$. The change in $A_y$ for a change from $a_{y1}$ to $a_{y2}$ is:

$$\Delta A_y = A_{y2} - A_{y1} = \frac{a_{y2}}{1 + a_{y2}f_z} - \frac{a_{y1}}{1 + a_{y1}f_z}$$

where $f_z$ is assumed to remain constant, independent of $a_y$. If we rearrange this equation and recall the relation $A_{y1} = a_{y1}/(1 + a_{y1}f_z)$, we obtain the desensitivity expression for large changes:

$$\frac{\Delta A_y}{A_{y1}} = \left(\frac{1}{1 + a_{y2}f_z}\right)\frac{\Delta a_y}{a_{y1}} \qquad (3.20)$$

For a numerical case, assume in the example above that $\beta_0$ increases by 50%; that is, $\beta_0$ increases from 50 to 75. Because $r_\pi$ and $r_x$ are masked by $R_e$ and $R_b$, $a_y$ increases from $21 \times 10^{-3}$ to $31 \times 10^{-3}$. So $a_{y2} = 31 \times 10^{-3}$ mhos in Eq. 3.20. With $f_z = 500$ ohms, we expect that the fractional change in $A_y$ is $(50/16.5)\% = 3.0\%$, even for such a large change of $\beta_0$.

### 3.3.3 *Example of Current Ratio Feedback*

The configuration shown in Fig. 3.14 samples $I_e$ at the output and provides a feedback current at the input. Let us limit consideration to frequencies at which $I_e \approx -I_L$. Analysis in feedback

### Sec. 3.3 Examples of Elementary Feedback Analysis

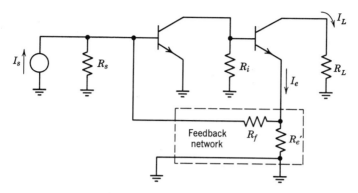

**Fig. 3.14.** A current-ratio feedback pair (approximately).

form would then require $V_{in}$ and $I_{out}$ as *independent* variables (which are the same for both the amplifier and the feedback network) and $I_{in}$ and $V_{out}$ as *dependent* variables. This set is not a particularly common one. Nonetheless, a little thought about steps analogous to those we employed for the transadmittance or transimpedance cases will make it clear that the basic gain should be taken as that defined by $a_i = I_L'/I_s$ in Fig. 3.15a, and the loop transmission as $I_t/I_s$ in Fig. 3.15b. Thus the feedback function $f_i$ will be given by $I_t/I_L'$ in Fig. 3.15c, or

$$f_i = -\frac{G_f}{G_f + G_e'} \tag{3.21}$$

Observe that the open-loop amplifier as defined in Fig. 3.15a has within it a second transistor with a common-emitter series feedback element (transimpedance feedback), as in Fig. 3.13b. On that account, we may say that $R_e'$ in Fig. 3.14 does *two* things: it functions as part of an *over-all* feedback loop, as in Fig. 3.15b, and it provides *local* feedback to the second stage of the basic amplifier when the "over-all loop" is open (the over-all feedback removed) in the manner prescribed by Fig. 3.15a.

Of course, we may use the loop transmission defined above to calculate the system impedance levels at input and output, noting that they are *lowered* by the factor $1 + T$ on the input and *raised* by this factor on the output for this case (compare Fig. 3.14 with Figs. 3.6d and 3.9d).

**98**  Feedback Amplifier Concepts

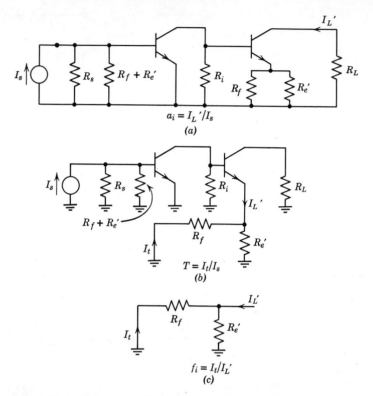

**Fig. 3.15.** Formulation of feedback analysis for the current-ratio feedback pair of Fig. 3.14. (*a*) Open-loop gain. (*b*) Loop transmission. (*c*) Feedback function.

Once the ideas expressed in Fig. 3.15 are appreciated, based as they are on the assumption that $I_e = -I_L$ in Fig. 3.14, the rest of the analysis is straightforward and need not be carried further here.

**PROBLEMS**

**P3.1**  Repeat Problem P2.7(*e*), except obtain the gain by adding negative feedback as shown in Fig. 3.16. Assume that $R_f$ is much greater than 200 ohms, so that $R_f$ does not load the output or input circuits significantly. Determine the proper value for $R_f$ to achieve the desired over-all gain. Make reasonable approximations in the analysis.

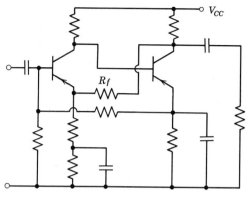

Fig. 3.16.

**P3.2** The circuit of a two-stage transistor amplifier with feedback is shown in Fig. 3.17.

(a) Show that with the feedback removed ($R_f = \infty$) this amplifier can be represented by the model shown in Fig. 3.18. Assume that signal voltage drops across coupling and bypass capacitors are negligible and that the signal current in the inductor $L$ is negligible. Estimate the quiescent collector currents, then estimate the midband values of the input resistance $R_i$ and the over-all transconductance $G_m$. Use the following transistor parameters: $\beta_0 = g_m r_\pi = 50$ and $r_x = 50$ ohms.

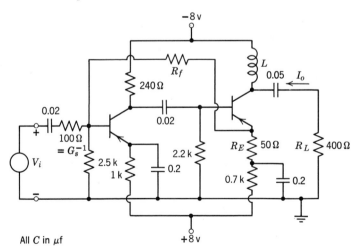

Fig. 3.17.

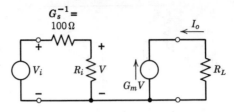

Fig. 3.18.

(b) If the feedback resistor is not too small, the amplifier with feedback can be represented by the model in Fig. 3.19. Determine $f_i$ in terms of $R_f$ and any other necessary parameters of the circuit. What is the minimum value of $R_f$ for which you would be willing to employ this model?

(c) Determine the mid-band value of the gain $I_o/G_s V_i$. Show that this gain can be put in the standard form $a/(1 + af)$ where $a$ is the open loop gain and $f$ is the feedback function.

**P3.3** A two-stage amplifier and its approximate low frequency, incremental model are shown in Fig. 3.20.
  (a) Estimate the voltage gain $V_L/V_s$ on the basis of physical reasoning.
  (b) Calculate the current gain $I_L/I_s$ on the basis of the model in (b). Use this calculation to determine the voltage gain accurately.
  (c) If a 1µf capacitor is connected across $R_L$, calculate the resulting natural frequency by finding the resistance facing the capacitor.

**P3.4** The circuit in Fig. 3.21 shows a three-stage transistor amplifier which uses feedback biasing. *Assuming that the amplifier is stable* and that the transistors have $\beta_F = \beta_0 = 50$,
  (a) Choose $R$ so that the collector-to-ground voltage ($V_C$) of the output transistor is 8 volts. What are the operating points ($I_C$, $V_{CE}$) of the transistors under this condition?
  (b) With $R$ fixed at the value determined in (a), the supply voltage changes to +16 volts. How does this change affect the operating points of the three stages?
  (c) What is the resistance seen by the capacitor $C$?

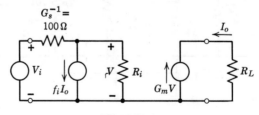

Fig. 3.19.

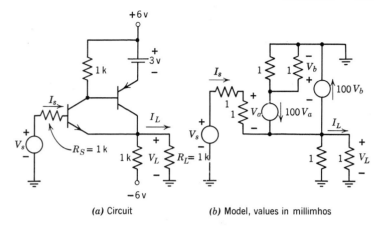

(a) Circuit        (b) Model, values in millimhos

Fig. 3.20.

(d) Choose $C$ so that the amplifier has a lower half-power frequency of 100 cps.

**P3.5** This problem and Problems P4.11 and P6.2 are concerned with the use of "operational amplifiers" in conjunction with various feedback connections. An operational amplifier is a high-gain, direct-coupled amplifier with a differential input connection. A circuit which is indicative of practical design is shown in Fig. 6.24 (page 222), and a simplified linear, incremental model is shown in Fig. 3.22a. The amplifier is assumed to be unilateral with $z_i \gg z_o$ and $a_v \gg 1$, and all extraneous sources within the

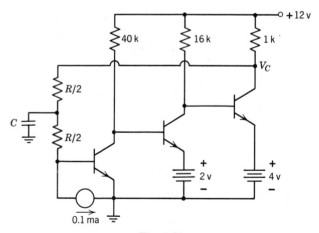

Fig. 3.21.

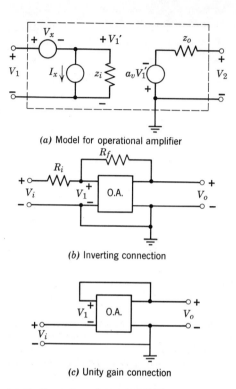

**Fig. 3.22.** Operational amplifier connections.

amplifier are represented by the $V_x$ and $I_x$ generators in Fig. 3.22a.

(a) Figure 3.22b shows one common connection which is called an "inverting" connection because a plus input produces a minus output and vice versa (i.e., the gain is negative). Assume that with $V_x = I_x = 0$ we have:

$$\frac{V_o}{V_i}(s=0) \equiv A_0 \quad \text{with } A_0 \ll a_v$$

If $A_0$ is specified, what values of $R_i$ and $R_f$ would you use to minimize the fractional change in $A_0$ per unit fractional change in $a_v$?

(b) If $V_x$ and $I_x$ are uncorrelated noise generators and $A_0$ is specified, what values of $R_i$ and $R_o$ would you use for minimum noise in the output? Assume $V_i$ is supplied from a voltage source and $R_i$ and $R_f$ contribute no noise.

(c) What are the approximate input and output resistances for the inverting connection?

(d) Fig. 3.22c shows a connection which provides very nearly unity gain. How close to unity is the open-circuit voltage gain?

**P3.6** In Fig. 3.23a appears a circuit diagram for a feedback amplifier (with dc bias supplies omitted).
(a) Show that for the behavior between terminals $b$–$b'$ and $c$–$c'$, Fig. 3.23b is approximately equivalent to Fig. 3.23a.
(b) Calculate $V_o/V_s$ when $R_{f2} = \infty$. Identify the open-loop gain and the loop transmission.
(c) Determine the input and output impedances at $b$–$b'$ and $c$–$c'$ for $R_{f2} = \infty$.
(d) Determine the effect of $R_{f2}$ on the answers to the preceding parts. In particular, discuss use of this scheme to control impedance levels precisely, in addition to providing other negative feedback advantages.

**P3.7** Determine the output impedance of the feedback amplifier of Fig. 3.12 by setting $G_L = 0$. In calculating $T_{G_L = 0}$, can the loading effects of $R_f$, $g_o$, and $g_\mu$ be neglected? If so, explain why. If not, explain why the answer (found by another method in the text) seems to be independent of these effects.

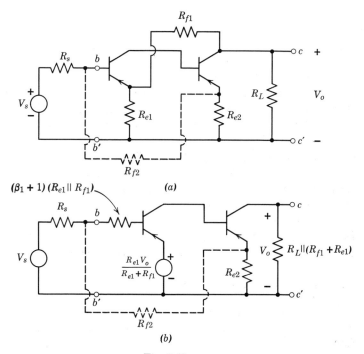

Fig. 3.23.

# 4

# Stability and Frequency Response of Feedback Amplifiers

## 4.0 INTRODUCTION

In Chapter 3 we discussed in some detail the feedback point of view and emphasized the desensitization to certain amplifier shortcomings that can be achieved through use of inverse feedback. It was shown that a sacrifice of gain is always one price of this desensitization; but no restrictions were placed on the amount of gain that may be sacrificed, i.e., upon the amount of inverse feedback that may be applied around a given original amplifier. Such restrictions will be taken up now. As suggested briefly at the end of the example of Sec. 3.3.1, the culprit in this respect is the frequency response of the basic amplifier (and the feedback network). It was pointed out that instability (the occurrence of unwanted oscillations) is the price paid for "excessive" feedback.

## 4.1 THE STABILITY PROBLEM

To examine the stability problem in more detail, let us study the locus of the natural frequencies of several feedback amplifier gain functions as we vary the amount of feedback. Throughout this chapter, we assume that *in the mid-frequency region of the amplifier we have negative feedback* although as we shall see, the feedback may well be positive at very high or very low frequencies. Also, *we assume that the basic amplifier is stable.*

Suppose, first, that the basic amplifier in Fig. 4.1 has a single pole at $s = s_a$, and that this pole is on the negative real axis, i.e., $s_a$ is real and negative:

$$a(s) = \frac{a(0)}{1 - s/s_a} \quad (4.1)$$

If we assume that the feedback function is independent of frequency, i.e., $f(s) = f_0$, then the closed-loop gain $A(s)$ of the feedback amplifier is:

$$A(s) = \frac{a}{1 + af} \quad (4.2a)$$

$$= \frac{a(0)}{1 + f_0 a(0) - s/s_a} \quad (4.2b)$$

The feedback amplifier gain function thus has one pole at

$$s = s_a[1 + f_0 a(0)] \quad (4.3)$$

Because the mid-band (here low-frequency) feedback is negative, that is, because $f_0 a(0) > 0$, the feedback has *increased* the upper 0.707 frequency of the amplifier by a factor $[1 + f_0 a(0)]$ and *decreased* the low-frequency gain by the same factor.

Note, however, that the feedback is fully effective in desensitizing the amplifier to parameter variations *only* when the loop trans-

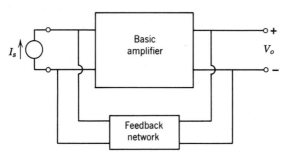

**Fig. 4.1.** Block diagram of amplifier with transadmittance feedback.

mission or loop gain is high. Hence, as the frequency is increased above $\omega = |s_a|$, the feedback becomes less effective. It is clear from Eq. 4.3 that this amplifier has no stability problems because it is not possible to move the pole of $A(s)$ into the right half-plane for any *positive* value of the low-frequency loop transmission or loop gain $f_0 a(0)$.

Consider next a basic amplifier with poles at $s = s_a$ and $s = s_b$, and assume again that the poles are on the negative real axis (i.e., $s_a$ and $s_b$ are negative real numbers):

$$a(s) = \frac{a(0)}{(1 - s/s_a)(1 - s/s_b)} = \frac{a(0)}{1 + a_1 s + a_2 s^2} \quad (4.4)$$

We now find for the closed loop gain, with the constant feedback function $f_0$:

$$A(s) = \frac{a(0)}{[1 + T(0)] + a_1 s + a_2 s^2} \quad (4.5)$$

where, for convenience, we have defined the loop transmission or loop gain as:

$$T(s) \equiv f(s) a(s) \quad (4.6)$$

and hence for this case at low frequencies

$$T(0) \equiv f_0 a(0)$$

From the denominator of Eq. 4.5, we find that $s_1$ and $s_2$, the poles of $A(s)$, lie on the negative real axis for $T(0)$ less than $(a_1^2/4a_2) - 1$. For $T(0)$ greater than this value, $A(s)$ has a complex pole pair. Figure 4.2a shows the locus of the poles of $A(s)$ for increasing $T(0)$. For convenience the pole locations have been expressed in terms of the $Q$ of the circuit by rewriting Eq. 4.5 as:

$$A(s) = \frac{A(0)}{1 + (1/Q)(s/s_o) + (s/s_o)^2} \quad (4.7)$$

where

$$s_o^2 = \frac{1 + T(0)}{a_2} \quad (4.8)$$

$$Q^2 = \frac{a_2[1 + T(0)]}{a_1^2} \quad (4.9)$$

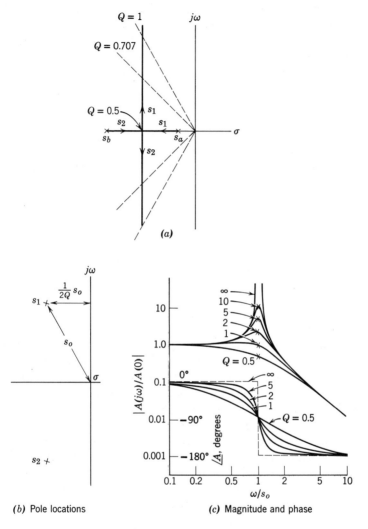

**Fig. 4.2.** (a) Loci of natural frequencies $s_1$ and $s_2$ of $A(s)$ as $T(0)$ is increased from zero in Eqs. 4.5 and 4.7. (b) Geometric interpretation of $Q$. (c) Normalized frequency response of $A(j\omega)$. In this chapter, poles of $T(s)$ and poles of $a(s)$ are designated by letter subscripts: $s_a$, $s_b$, $s_c$; whereas poles of $A(s)$ are designated by numbers: $s_1$, $s_2$, $s_3$.

The geometric interpretation of $Q$ and $s_o$ is given in Fig. 4.2b. The *normalized* frequency response of $A(j\omega)$ for various values of $Q$ is shown in Fig. 4.2c. The peak value of normalized gain is slightly greater than $Q$ and occurs for $\omega$ slightly less than $s_o$. If $Q$ is high, however, we can usually assume the peak occurs at $\omega = s_o$ and has a value $Q$. Note that the poles of $A(s)$ always lie in the left half-plane, and hence again there is no stability problem. However, if the $Q$ is much greater than unity, the amplifier will have a markedly peaked amplitude response and a damped oscillation in the step response.

As a third example of a simple feedback amplifier, consider the case where the basic amplifier has three poles on the negative real axis:

$$a(s) = \frac{a(0)}{(1 - s/s_a)(1 - s/s_b)(1 - s/s_c)} \quad (4.10)$$

For $f(s) = f_0$, we have

$$A(s) = \frac{a(0)}{[1 + T(0)] + a_1 s + a_2 s^2 + a_3 s^3} \quad (4.11)$$

This function always has one real pole, but for large enough values of $T(0)$ the other two poles can break away from the real axis to form a complex pair. In addition, for sufficiently large values of $[1 + T(0)]$, *this complex pair moves into the right half-plane.* That is, the amplifier can become *unstable*.

Typical loci of poles of $A(s)$ for Eq. 4.11 are shown in Fig. 4.3. The frequency response and step response of the amplifier can be envisioned to some extent from the $Q$ of the complex pair of poles. However, the real axis pole tends to smooth out the hump in the frequency response and the ringing in the step response which would result from the complex pair alone.

### 4.2 ROOT LOCUS CALCULATIONS

#### 4.2.0 *Introduction*

It should be clear from the preceding three examples that the amount of loop gain $T(0)$ in a feedback amplifier can profoundly

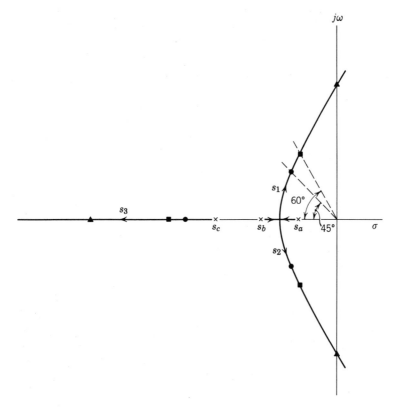

**Fig. 4.3.** Loci of natural frequencies $s_1$, $s_2$, and $s_3$ of $A(s)$ as $T(0)$ is increased from zero. Also shown are pole locations for four specific values of $T(0)$:

$\times\ T(0) = 0$
$\bullet\ T(0) = 2.4$
$\blacksquare\ T(0) = 4.4$
$\blacktriangle\ T(0) = 31$

assuming poles of $T(s)$ are at $s_a = -0.1$, $s_b = -0.7$, and $s_c = -1.8$. (Figure not to scale)

influence the location of the amplifier natural frequencies, and hence its frequency response and transient response. An important problem in feedback amplifier design, then, is how to choose a basic amplifier and feedback network combination to obtain at the same time both large loop gain (and hence good desensitivity), and acceptable frequency response and transient response. There are two principal methods of analysis which are helpful in attacking this problem. One method concentrates on the location of the natural frequencies of $A(s)$ in the complex plane — this is the root-locus method (essentially the method we used in the preceding section to introduce the stability problem). We discuss this method in more detail in this section. A second method, which concentrates on the sinusoidal steady-state response $A(j\omega)$ of the amplifier for various loop gain functions $T(j\omega)$, will be discussed in Sec. 4.3.

The essential idea of the root-locus method is to trace out the path of the feedback amplifier natural frequencies in the $s$ plane, as a function of the mid-frequency loop gain $T(0)$. On the basis of this locus, compensating networks can be added either to the basic amplifier or the feedback network in order to reshape the locus to improve amplifier performance. That is, we can reshape the locus to improve the pole locations for a given amount of midband feedback.

The relationships between the pole (and zero) locations of an amplifier and its frequency response and transient response are not simple. We do know, however, that when the amplifier has complex poles, it is likely that the transient response will be oscillatory and the frequency response will be peaked at the ringing frequency. Specifically, for a single pair of complex poles with a $Q$ greater than 0.5 (i.e., lying off the axis) there will be ringing in the step response. For a $Q$ greater than 0.707 (45° radials) the frequency response will be peaked as well (see Fig. 4.2). The requirement on pole locations for "satisfactory" response depends principally on the application. For example, in an oscilloscope amplifier, a ringing step response is intolerable, so we require that the $Q$ of complex poles be close to 0.5. On the other hand, for audio amplifiers, some small peaking in the frequency response might be permitted, so the $Q$ might be as high as 1. In all but the simplest of designs (a one-stage amplifier, for example) these requirements will force us to use some type of compensation in order to achieve a reasonable amount of loop transmission or loop gain.

It is clear from the examples which we worked in Sec. 4.1 that finding the exact root locus for a given amplifier could involve a fair amount of calculation. Thus, before proceeding to a specific example to illustrate the root-locus method, we first develop a number of simple rules for sketching approximate loci with a minimum of computation.

### 4.2.1 Construction of Approximate Loci

The expression for the closed-loop gain is

$$A(s) = \frac{a(s)}{1 + T(s)} \quad (4.12)$$

Instead of factoring the polynomial $1 + T(s)$ for several values of $T(0)$ to find the loci of the poles of $A(s)$, we can establish some simple guides which permit us to sketch the loci with reasonable accuracy.

1. *Start and End of Locus.* The first such guide is that the poles of $A(s)$ for increasing amounts of feedback move from the poles of $T(s)$ to the zeros of $T(s)$. To show this, we express the loop gain $T(s)$ as

$$T(s) = T(0)g(s) \quad (4.13)$$

where

$$T(0) = a(0)f(0) \quad (4.14)$$

and $g(s)$ is a ratio of polynomials which is by definition equal to unity at mid-band. Hence

$$A(s) = \frac{a(s)}{1 + T(0)g(s)} \quad (4.15)$$

For finite and nonzero values of $T(0)$, the poles of $A(s)$ occur *only* where $T(0)g(s) = -1$. If $T(0)$ is very small, then at a pole of $A(s)$, $g(s)$ must be very much *larger* than one, i.e., we must be close to a pole of $g(s)$ and hence a pole of $T(s)$. Conversely, if $T(0)$ is very large, then at a pole of $A(s)$, $g(s)$ must be much *less* than one, i.e., we must be close to a zero of $T(s)$. Thus we can say that *for increasing amounts of midband feedback, the locus of each pole of $A(s)$ starts from a pole of $T(s)$ and terminates on a zero of $T(s)$.* If $T(s)$ has more finite poles than finite zeros, some of the loci will terminate on the zeros at infinity. [Notice that the poles of $a(s)$ appear in $g(s)$ also, as shown by Eqs. 4.6 and 4.13. Therefore, as stated above, as long as $0 < T(0) < \infty$, $A(s)$ does *not* have the

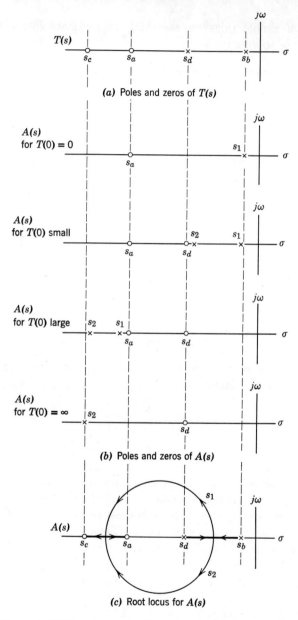

**Fig. 4.4.** Motion of $s_1$ and $s_2$, the poles of $A(s)$, for increasing $T(0)$.

Sec. 4.2 Root Locus Calculations    113

poles of $a(s)$. However, when $T(0) = 0$, $A(s)$ has poles *only* at the poles of $a(s)$. Moreover, for $T(0) = \infty$, $A(s)$ has poles *only* at the zeros of $f(s)$].

As a simple example, consider the case where the basic amplifier has a zero at $s_a$ and a pole at $s_b$, and the feedback network has a zero at $s_c$ and a pole at $s_d$, as shown in Fig. 4.4a. The poles of $A(s)$ for $T(0)$ zero, small, large, and infinite are shown in Fig. 4.4b.

2. *Location of Loci on Real Axis.* As a second guide to drawing the loci, we can easily locate those portions of the loci of the poles of $A(s)$ which lie on the real axis. At a pole of $A(s)$, $1 + T(s) = 0$, so $|T(s)| = 1$ and $\angle T(s) = \pi$. But on the real axis, $\angle T(s)$ can only be zero or $\pi$. Consider the example shown in Fig. 4.5. At a point $A$, the phase contributions of the real poles and real zeros of $T(s)$ are zero, as shown by the angle of the vectors from the poles and zeros to the point. The phase contributions of the complex pair cancel out everywhere on the real axis. The phase of $T(s)$ for this value of $s$ is zero.

For a point $B$, which lies between $s_a$ and $s_b$, $\angle T(s) = \pi$. For a point between $s_b$ and $s_c$, $\angle T(s) = 0$. We can generalize from this example, and the fact that the poles of $A(s)$ move *continuously* with $T(0)$, to obtain the rule: *The loci of the poles of $A(s)$ include all portions of the real axis to the left of an odd number of poles and zeros of $T(s)$.* (Because $T(0) > 0$, right-half-plane zeros are disregarded.)

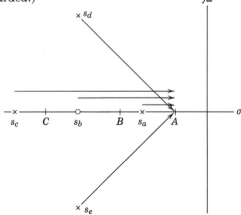

**Fig. 4.5.** Calculation of locus on real axis from the poles and zeros of $T(s)$.

3. *Formation of a Complex Pole Pair.* Figure 4.4 represents an interesting illustration of Rule 2. For this case Rule 2 says that the real-axis loci of the poles of $A(s)$ must lie only between $s_a$ and $s_c$, and between $s_b$ and $s_d$, as shown in Fig. 4.4c. But we know from Rule 1 that for $T(0)$ small and increasing, one of the poles of $A(s)$ starts at $s_b$ and the other at $s_d$. Similarly, when $T(0)$ becomes large, one pole of $A(s)$ terminates at $s_a$ and the other at $s_c$. In order to move continuously from $s_b$ and $s_d$ to $s_a$ and $s_c$, and at the same time not violate the condition on real-axis loci, the poles of $A(s)$ must leave the real axis and form a complex pole pair, as shown in Fig. 4.4c. The example in Fig. 4.4 thus brings out another guide to locus construction: *All segments of loci which lie on the real axis between pairs of poles or pairs of zeros of $T(s)$ must at some internal "breakaway point" branch out from the real axis into conjugate pairs of loci.*

4. *Center of Gravity of Poles.* Equation 4.11 serves to illustrate another guide. Recall from Eqs. 1.41 and 1.62b that:

$$\frac{a_{n-1}}{a_n} = -\sum s_j \qquad (4.16)$$

Thus in situations such as Eq. 4.11, where the two highest coefficients of the denominator polynomial are not functions of $T(0)$, $\sum s_j$ must remain constant. It is sometimes more convenient to phrase this in terms of the "center of gravity" of the poles, by analogy with the calculation of the center of gravity of a number of point masses in mechanics. To this end we divide both sides of Eq. 4.16 by the number of poles, thus:

$$-\frac{a_{n-1}}{na_n} = \frac{\sum s_j}{n} = \text{center of gravity of poles} \qquad (4.17)$$

Thus, *if the coefficients $a_n$ and $a_{n-1}$ of $A(s)$ are independent of $T(0)$, the center of gravity of the poles of $A(s)$ will be constant, equal to the center of gravity of the poles of $T(s)$.* The coefficients $a_n$ and $a_{n-1}$ will be independent of $T(0)$ if there are at least two more finite poles than finite zeros in $T(s)$.

5. *Pole Locations for Cubic Denominator.* The gain expression with a cubic denominator in the form of Eq. 4.11 is one which is frequently encountered in feedback calculations. Rather than solv-

ing directly to find the pole locus as a function of $T(0)$, it is easier to find the locus by assuming the pole angles, and calculating the value of $T(0)$ required to meet these constraints. For the cubic case, we assume that $A(s)$ has one real pole, and a pair of complex poles at a specified angle from the real axis (or equivalently, of a specified $Q$). We then solve for $T(0)$ by equating the coefficients $b_3$, $b_2$, $b_1$, and $b_0$ in

$$A(s) = \frac{k}{b_3 s^3 + b_2 s^2 + b_1 s + b_0}$$

to the coefficients of a cubic polynomial with a pair of complex roots at the desired angle. The results are shown in Table 4.1.

**TABLE 4.1 Root locus for cubic denominator.**

| $Q$ of Complex Pair | Angle from Real Axis | $\alpha$ Real Part of Complex Pole | $\gamma$ Real Pole | $b_0$ |
|---|---|---|---|---|
| 0.5 | 0° (two poles together) | $\left(\alpha^2 + \frac{2}{3}\frac{b_2}{b_3}\alpha + \frac{b_1}{3b_3} = 0\right)$ | $-\frac{b_2}{b_3} - 2\alpha$ | $-\alpha^2 \gamma b_3$ |
| 0.707 | ±45° | $\left(\alpha^2 + \frac{b_2}{b_3}\alpha + \frac{b_1}{2b_3} = 0\right)$ | $-\frac{b_2}{b_3} - 2\alpha$ | $-2\alpha^2 \gamma b_3$ |
| 1 | ±60° | $-\frac{b_1}{2b_2}$ | $-\frac{b_2}{b_3} - 2\alpha$ | $-4\alpha^2 \gamma b_3$ |
| ∞ | ±90° | 0 | $-\frac{b_2}{b_3}$ | $\frac{b_1 b_2}{b_3}$ |

Given any three of $b_3$, $b_2$, $b_1$, $b_0$, $\gamma$, and $\alpha$ (or appropriate interrelations), and either the $Q$ or the angle of the poles, we can use Table 4.1 to find the remaining parameters.

### 4.2.2 Example of Root Locus Technique

As an example of root-locus calculations, let us study a simple three-stage amplifier with transadmittance feedback, as in Fig. 4.6. In the following sections, we discuss circuit modifications of the

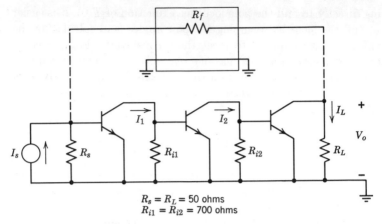

**Fig. 4.6.** Circuit of amplifier with transadmittance feedback. Biasing networks and power supplies have been omitted.

basic amplifier and the feedback network which make it possible to obtain at the same time both large loop gain (and hence desensitivity), and an acceptable amplifier response. The desensitivity properties of this circuit were discussed in Sec. 3.3.1 and need not be repeated here.

The open-loop gain $a(s)$ for the amplifier will be of the form:

$$a(s) = \frac{a(0)}{a_3 s^3 + a_2 s^2 + a_1 s + 1} \quad (4.18)$$

where for this configuration $a(0)$ will be a *negative* number. In this equation we have included only the *dominant* natural frequencies associated with the high-frequency behavior. There are in fact three more poles at very high frequencies, the effects of which we shall discuss later in this chapter. Also, we have not included any of the low-frequency poles and zeros resulting from the coupling and bypass capacitors normally present in an ac amplifier. That is, we have assumed that the low-frequency gain is constant down to $s = 0$ at a value $a(0)$. This procedure is followed only for simplicity. It should be emphasized that the low-frequency stability problem in a feedback amplifier can be just as difficult as the high-frequency stability problem. However, the method of analysis is identical, so only the high-frequency problem will be discussed here.

The feedback function for this amplifier is:

$$f = y_{rf} = -G_f \tag{4.19}$$

Thus the loop gain is:

$$T(s) = a(s)f(s) = \frac{T(0)}{a_3 s^3 + a_2 s^2 + a_1 s + 1} \tag{4.20}$$

where $T(0) = -a(0)G_f$, a *positive* number. The closed-loop gain function, from Eq. 4.2a, is:

$$A(s) = \frac{a(0)}{a_3 s^3 + a_2 s^2 + a_1 s + [1 + T(0)]} \tag{4.21}$$

The general form of the root locus for Eq. 4.21 is shown in Fig. 4.3.

To make the example more specific, let us assume that the dominant natural frequencies of the basic amplifier, that is, the dominant poles of $a(s)$, are:

$$s_a = -0.1 \times 10^8 \text{ sec}^{-1}$$

$$s_b = -0.7 \times 10^8 \text{ sec}^{-1}$$

$$s_c = -1.8 \times 10^8 \text{ sec}^{-1}$$

(These correspond approximately to the three dominant poles calculated for the amplifier in Chapters 1 and 2.) We know from Sec. 4.2.1 that the loci of the poles of $A(s)$ start from $s_a$, $s_b$, and $s_c$. On the real axis the locus can only be between $s_a$ and $s_b$, and to the left of $s_c$. Also, the center of gravity of the three poles of $A(s)$ is:

$$-\frac{a_{n-1}}{n a_n} = \frac{\sum s_j}{3} = \frac{-2.6 \times 10^8}{3} = -0.866 \times 10^8 \text{ sec}^{-1}$$

On the basis of the numerical values given above, the closed-loop gain expression becomes:

$$A(s) = \frac{a(0)}{8 s^3 + 20.7 s^2 + 12 s + 1 + T(0)} \tag{4.22}$$
[$s$ in units of $(\text{sec})^{-1} \times 10^8$]

To keep the $Q$ of the complex pole pair less than 0.707, we find from Table 4.1 that the mid-frequency loop gain $T(0)$ must be less than 2.4. For this particular pole configuration, acceptable response will still result even if the poles approach the $\pm 60°$ radials ($Q = 1$), because of the influence of the real pole. However, even under this condition $T(0)$ still cannot be greater than 4.4, a relatively small amount of feedback. With $T(0) = 4.4$, the poles of $A(s)$ are located at $s_1, s_2 = (-0.29 \pm j\,0.5) \times 10^8$ sec$^{-1}$, $s_3 = -2 \times 10^8$ sec$^{-1}$. Thus the bandwidth of the amplifier will be, from Fig. 4.2, about $0.75 \times 10^8$ rad/sec.

### 4.2.3 *Modifications of the Basic Amplifier*

Clearly, we need a substantially greater amount of feedback than $T(0) = 4.4$ if we wish to obtain a reasonable amount of amplifier desensitivity. In this and the following subsection we examine various ways of *compensating* the loop gain $T(s)$ so that we can increase $T(0)$ while still keeping the $Q$ of the complex pole pair less than 1. One way of accomplishing this is to move the lowest pole of $T(s)$ further in on the negative real axis.

Of course, in a transistor amplifier we usually do not have independent control over the pole locations, so it is more helpful to phrase the discussion in terms of the coefficients of the denominator polynomial in Eq. 4.21. We see from Table 4.1 that to prevent oscillations (i.e., keep $Q$ less than infinity),

$$T(0) < \left(\frac{a_1 a_2}{a_3} - 1\right) \tag{4.23}$$

Thus, to achieve higher loop gain without oscillations, we can increase $a_1$ or $a_2$, or decrease $a_3$, in Eq. 4.21. We can correctly assume that the same changes in $a_1$, $a_2$, or $a_3$ will also result in a corresponding increase in the allowable loop gain for a $Q$ of 1.

We can increase $a_1$ in Eq. 4.21 by raising the value of the sum of the open-circuit time constants *in the basic amplifier*. That is, in Eq. 4.18, we increase

$$\sum \tau_{jo} = \sum R_{jo} C_j = a_1 \tag{4.24}$$

(See Eq. 1.39, but set $a_0 = 1$.) The sum of the time constants can be increased by increasing any one of the open-circuit resistances $R_{jo}$; for example, by increasing $R_c$, or adding a resistor in series with the collector, as shown in Fig. 4.7. The operating point

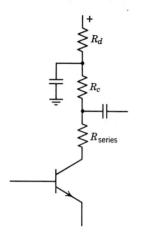

Fig. 4.7. Increasing $\sum R_{jo}C_j$ by adding a resistor in the collector circuit.

for the transistor can be maintained at its previous value by reducing $R_d$. These circuit changes in $a(s)$ will, in fact, change not only $a_1$, but $a_2$ and $a_3$ as well. Recall from Eq. 1.41 that

$$\frac{a_2}{a_3} = \sum \frac{1}{\tau_{js}} \qquad (4.25)$$

But increasing $R_c$, for example, *must* increase $\tau_{js}$ as well as $\tau_{jo}$, and thus decrease $a_2/a_3$. An increase in the allowable $T(0)$ will be realized only if we add resistance in such a way as to increase the open-circuit time constant $\sum \tau_{jo}$ as much as possible, and decrease $\sum (1/\tau_{js})$ as little as possible (see Eq. 4.23).

We can also increase $\sum R_{jo}C_j$ by adding capacitance, either between base and ground, or between base and collector, as shown by the dotted lines in Fig. 4.8. The capacitors $C_1$ and $C_2$ will have much the same effect as increasing $C_\pi$ and $C_\mu$, respectively. Therefore, the dominant effect will be to move one or more poles closer in toward the origin. The advantage of adding $C_2$ instead of $C_1$ is that adding $C_2$ tends to split the poles apart, i.e., not only move $s_1$ in, but also move $s_3$ out, as we saw in Chapter 1.

Note that obtaining a large value of $T(0)$ by increasing $a_1$ has important disadvantages. We see from Eq. 1.17 that for the basic

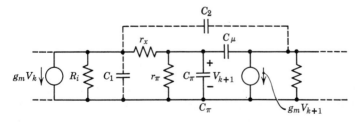

Fig. 4.8. Increasing $\sum R_{jo}C_j$ by adding capacitance $C_1$ or $C_2$.

amplifier, and hence for the loop gain $T(s)$, the upper 0.707 point will be

$$\omega_h \cong \frac{a_0}{a_1} = \frac{1}{a_1} \qquad (4.26)$$

because, from Eq. 4.18, $a_0 = 1$ for this example. Increasing $a_1$ has thus reduced the pass band of $T(j\omega)$. *Therefore, the desensitization of the amplifier is effective over a much smaller passband.*

### 4.2.4 Modification of the Feedback Network

A second important way of compensating $T(s)$ to permit more mid-frequency loop gain is to modify the feedback network by introducing reactive elements. In this section, we take up only the simple feedback network shown in Fig. 4.9 because, as we shall see, the one added capacitance $C_f$ gives us in effect independent control of the coefficient of the $s$ term in the denominator of $A(s)$. Thus extensive modification of the root locus is possible.

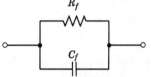

**Fig. 4.9.** Feedback network.

For the feedback network, the reverse transfer admittance is:

$$y_{rf} = \frac{-1}{R_f}(1 + sR_fC_f) = f(s) \qquad (4.27)$$

The loop gain, $T(s)$, now has a finite zero:

$$T(s) = af = T(0)\frac{(1 - s/s_f)}{(1 - s/s_a)(1 - s/s_b)(1 - s/s_c)} \qquad (4.28)$$

where

$$s_f = -\frac{1}{R_fC_f} \qquad (4.29)$$

and where it is assumed that $a(s)$ still provides three poles and no finite zeros as in Secs. 4.2.2 and 4.2.3. (We are thus *neglecting the loading* of the feedback network on the *input and output* of the amplifier.)

At this point it is instructive to form $A(s)$, the gain function for the complete amplifier, to see how the zero of $f(s)$ affects the gain:

$$A(s) = \frac{a(s)}{1 + T(s)}$$

$$= \frac{a(0)}{(1 - s/s_a)(1 - s/s_b)(1 - s/s_c) + T(0)(1- s/s_f)} \quad (4.30a)$$

$$= \frac{a(0)}{a_3 s^3 + a_2 s^2 + [a_1 + R_f C_f T(0)]s + 1 + T(0)} \quad (4.30b)$$

Equation 4.30b indicates that by adding capacitor $C_f$, we now have independent control of the coefficient of the $s$ term in the denominator of $A(s)$. Thus, in accordance with the discussion in the preceding section, we can by this method achieve large amounts of loop gain and hence good desensitivity.

(The thoughtful reader may well ask why, when we consider the *feedback amplifier as a whole*, $C_f$ contributes to $\sum \tau_{jo}$ in Eq. 4.30b, but because $a_2$ and $a_3$ are unchanged, it apparently does not contribute to $\sum (1/\tau_{js})$. The answer is that $C_f$ will, in fact, add a fourth pole to the system, and hence an $a_4 s^4$ term to the denominator of Eq. 4.30b. Thus $\sum (1/\tau_{js})$, now equal to $a_3/a_4$, will in fact change; but $a_2/a_3$ will not change appreciably, because the fourth pole will be so far out on the negative real axis. We have neglected this fourth pole by neglecting the loading of the basic amplifier by $C_f$.)

Returning now to Eq. 4.30a, we note that $A(s)$ does not have any finite zeros. The zero $s_f$ (or any finite zero of $f$ for the general case) does not appear directly in the over-all gain expression. However, $s_f$ certainly affects the natural frequencies of the total amplifier, as can be seen from the denominator of Eq. 4.30a. Our technique will be to place the zero of the feedback network in such a location as to modify the loci in a favorable manner.

Recall that the loci of the poles of $A(s)$ for the amplifier *without* $C_f$ are shown in Fig. 4.3. The corresponding loci *with* $C_f$ included are shown in Fig. 4.10. In forming the loci in Fig. 4.10, both $R_f$ and $C_f$ must be changed such that the product $R_f C_f$ stays constant. In this way we can vary $T(0)$ and at the same time keep the feedback zero $s_f$ fixed.

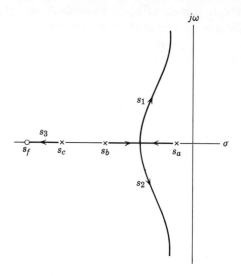

**Fig. 4.10.** Pole-zero locations of $T(s)$ and corresponding root locus for $A(s)$. (Figure not to scale.)

It is clear from Sec. 4.2.1 that one real-axis locus will start from $s_c$ and terminate on $s_f$. The other two poles will come together as before and form a complex pair. However, because in this case the center of gravity of the poles is unchanged by the feedback zero, and pole $s_3$ moves only to $s_f$, *the asymptotes for the complex pair for large $T(0)$ must be parallel to the $j\omega$ axis.* Furthermore, these loci, for a proper choice of $s_f$, never cross into the right half-plane. The feedback amplifier is then always stable (if only dominant effects are considered). The value of $s_f$ to insure that these asymptotes do not cross into the right half-plane can be found quite readily. We know that the center of gravity remains constant at $(-2.6/3) \times 10^8$ sec$^{-1}$. Thus, for $s_1$ and $s_2$ to end up on the $j\omega$ axis, $s_3$ must end up at $-2.6 \times 10^8$ sec$^{-1}$. Thus $s_f = -2.6 \times 10^8$ sec$^{-1}$. To move $s_1$ and $s_2$ to the left away from the $j\omega$ axis, clearly we need only move $s_f$ closer to the origin by increasing $C_f$.

There are obviously many ways of choosing the value of $s_f$. We could, for example, adjust the zero to coincide with one of the poles of $a(s)$, say $s_b$. On the basis of the center-of-gravity argument, the loci then become straight lines parallel to the $j\omega$ axis, as shown in

Fig. 4.11. The asymptotes for large $T(0)$ clearly lie in the left half-plane.

A more practical case might be to choose $s_f$ so that the three poles of $A(s)$ fall on a circle, with the complex poles on the $\pm 60°$ radials. This is the pole configuration of a *Butterworth* filter, which gives the maximum bandwidth without peaking for a given center of gravity of the three poles. Because the center of gravity must be $(-2.6/3) \times 10^8 \text{ sec}^{-1}$, we find

$$\text{Re}[s_1] = \text{Re}[s_2] = -0.65 \times 10^8 \text{ sec}^{-1}$$
$$s_3 = -1.3 \times 10^8 \text{ sec}^{-1}$$

Thus

$$-s_1 = 1.3 \angle -60° \times 10^8 \text{ sec}^{-1}$$
$$-s_2 = 1.3 \angle 60° \times 10^8 \text{ sec}^{-1}$$
$$-s_3 = 1.3 \angle 0° \times 10^8 \text{ sec}^{-1}$$

In Eq. 4.30b, the constant term $[1 + T(0)]/a_3$ is the negative of the product of the roots. Using the value of $a_3$ from Eq. 4.22, we find

$$1 + T(0) = (1.3 \times 10^8)^3 (8 \times 10^{-24})$$
$$T(0) = 16.6$$

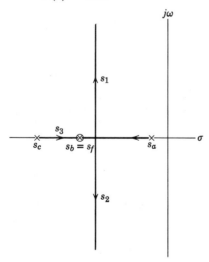

**Fig. 4.11.** Root locus for $A(s)$ with pole-zero cancellation. (Figure not to scale.)

For the Butterworth filter, $\omega_h = |s_3|$, so the 0.707 frequency of the feedback amplifier is $1.3 \times 10^8$ rad/sec. Comparison with the results in Sec. 4.2.2 indicates that addition of $C_f$ has brought about a substantial increase in both loop gain and bandwidth. Also, the Butterworth response achieved here will have no peaking, whereas the amplifier in Sec. 4.2.2 will have a slightly peaked response.

To find $R_f$ and $C_f$, we note that from Eq. 1.56b and Eq. 4.30b,

$$-\left(\frac{1}{s_1} + \frac{1}{s_2} + \frac{1}{s_3}\right) = \frac{a_1 + R_f C_f T(0)}{1 + T(0)}$$

Using the values for the roots as given above, and the value of $a_1$ from Eq. 4.22, we find

$$R_f C_f = 0.91 \times 10^{-8} \text{ sec}$$

From Sec. 3.3.1, $a(0) = -(26{,}000)(50)$. Therefore, because $T(0) = a(0)f(0) = -a(0)G_f$,

$$G_f = \frac{16.6}{(26{,}000)(50)} = 0.0127 \text{ mmhos}$$

Thus

$$C_f = (0.91 \times 10^{-8})(1.27 \times 10^{-5}) = 0.115 \text{ pf}$$

This capacitor value is impractically small, compared to the stray capacitance already present in the circuit. Thus it would be wise to redesign the feedback network by adding a 20:1 resistive voltage divider at the amplifier output, and connecting $R_f$ and $C_f$ to the tap on the divider. Then the feedback design would remain unchanged if both $G_f$ and $C_f$ were increased by a factor of 20.

To recapitulate, in this section we introduced a zero in the feedback network to modify the loci of the natural frequencies. For our example, we now can have a loop gain of 16.6, with a corresponding desensitivity of 17.6. Note in particular how simple it was to adjust the amplifier to have a prescribed pole pattern. This simplicity of design is one of the important advantages of the root locus approach.

In the cascaded basic amplifier, there should really be six poles of $a(s)$. However, we have used only the three dominant poles in all of the calculations in this section. The principal effect of the nondominant poles over the frequency range of interest here is to move the loci closer to the $j\omega$ axis. Thus the amplifier will be somewhat less stable than predicted from our approximate calculations. The effect of the nondominant poles is somewhat easier to see from the frequency response point of view to be presented in Sec. 4.3, so the matter will not be pursued further here.

### 4.2.5 Compensation by Adding Zeros to $a(s)$

In the preceding sections we have discussed in terms of an example two methods of compensating a feedback amplifier. First, we added either series resistance or shunt capacitance to the basic amplifier to move an already existing pole closer to the origin. Second, we made the feedback network frequency-dependent, thereby introducing a new zero into the loop gain $T(s)$ which brought about a major change in the root-locus plot.

A third technique for compensation is to introduce zeros in the transfer function $a(s)$ of the basic amplifier. If additional capacitance has already been added to the amplifier as in Fig. 4.8, then a zero can be added to $a(s)$ merely by adding a resistor in series with the capacitor ($C_1$ or $C_2$). This circuit produces a zero of $a(s)$ close to $s = -1/RC$, and also adds another pole to the system. The effect of the zero on the stability of the feedback amplifier is similar to that of a zero in the feedback network, as discussed in Sec. 4.2.4, except that the zero of $a(s)$ also appears as a zero of $A(s)$.

Zeros can also be added to $a(s)$ by using a series $RLC$ circuit for shunt loading between stages. The capacitor tends to move the lowest natural frequency of the basic amplifier nearer to the origin. The $RLC$ is then tuned to create a complex pair of zeros which can suppress the effect of high $Q$ poles in the feedback-amplifier response.

## 4.3 FEEDBACK AMPLIFIER DESIGN BASED ON j-AXIS RESPONSE

### 4.3.0 Introduction

We discussed in the preceding section methods of feedback amplifier design using root-locus plots. In this section we discuss the problem from a somewhat different point of view, in which the design is based solely on the behavior of the loop gain and the closed-loop amplifier gain on the $j\omega$ axis. That is, the design is based on the steady-state response $T(j\omega)$ and $A(j\omega)$. The two methods are of course closely related, and the circuit designer can very profitably use both concepts in a given design problem; for example, he can think conceptually in terms of root locus, while carrying out the detailed analysis of measured data in terms of the steady-state response.

In essence, the steady-state method involves finding the steady-state loop gain, $T(j\omega)$, either by calculation or measurement, and from this, calculating the steady-state response of the closed-loop amplifier, $A(j\omega)$. By direct analogy with the root-locus method, we then must decide if the design is satisfactory in terms of loop gain, $T(0)$ (and hence desensitivity), and frequency and transient response. If the design is not satisfactory, then again we must employ compensation by using the methods discussed in Sec. 4.2. The new loop-gain function, $T(j\omega)$, is then calculated, followed by the new $A(j\omega)$. An important advantage of this method of design is that it can be carried out *directly in terms of the measured frequency response of the basic amplifier.*

### 4.3.1 The Nyquist Criterion

It is essential to this method that we develop criteria for both stability and "satisfactory" transient response in terms of $T(j\omega)$ and $A(j\omega)$, just as we did in terms of poles and zeros in the root-locus method. We shall develop the criterion for stability in this section, and some criteria for satisfactory transient response in Sec. 4.3.2.

The basic stability criterion used in the root-locus discussion was that the natural frequencies of the feedback-amplifier transfer function had to be in the left half of the $s$-plane for the amplifier to be stable. We shall show that the corresponding characteristic of the steady-state response $T(j\omega)$ is that it must satisfy the Nyquist Criterion for amplifier stability:

A feedback amplifier will be unstable if a plot of the log magnitude of $T(j\omega)$ versus the phase of $T(j\omega)$ over the range $-\infty < \omega < +\infty$ encircles or passes through any of the points $T = -1$.

The criterion is based on the assumption that the function $T(s)$ itself is stable.

A plot of $\log |T(j\omega)|$ versus $\angle T(j\omega)$ for a typical amplifier is shown in Fig. 4.12a. The corresponding $s$-plane plot showing the locus is shown in Fig. 4.12b. The key idea of the Nyquist test is that *the $T(j\omega)$ locus plotted in Fig. 4.12a corresponds to the $j\omega$ axis in the $s$-plane, Fig. 4.12b*. The two plots are conformal, in the sense that small squares in one plane appear as small squares in the other, and 90° turns in one plane appear as 90° turns in the other. Unfortunately, the conventional way of plotting $\log |T(j\omega)|$ versus $\angle T(j\omega)$, as shown in Fig. 4.12a, is *reversed* from that required

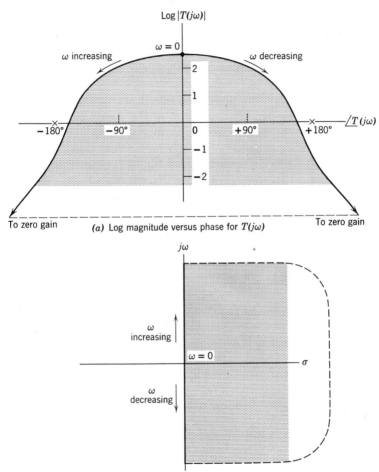

**Fig. 4.12.** Nyquist test.

**128** *Stability and Frequency Response of Feedback Amplifiers*

for conformality, because positive angle is plotted to the *right* instead of the *left*. Thus the angular relationships in Fig. 4.12 are *reversed*: $+90°$ turns in one plane correspond with $-90°$ turns in the other. In this example when we "walk" up the $j\omega$ axis in the $s$-plane in the direction of increasing $\omega$, the right half-plane always appears on our *right*. If we take a similar "walk" in the log magnitude versus angle plot in the direction of increasing $\omega$ along the corresponding path, that is, on the locus $T(j\omega)$, we must find the region corresponding to the right half of the $s$-plane on our *left*. To emphasize the point, these two corresponding regions have been shaded in the figure. Thus we have established that the right half-plane in the $s$-plane corresponds to the portion of the $T$ plot *inside* the $T(j\omega)$ locus.

The poles of $A(s)$ occur where $T = -1$, as we discussed at the start of this chapter. Thus, in terms of the log magnitude versus phase plots, the Nyquist Criterion requires that to insure the stability of $A(s)$, the points $T = -1$ must fall in that part of the plot corresponding to the left half of the $s$-plane. Specifically, *the points $T = -1$, that is, $|T| = 1$, $\angle T = \pm(2n + 1)180°$, must lie outside of the $T(j\omega)$ locus*. On this basis, the feedback amplifier in Fig. 4.12 is stable. (See Problem P4.5.)

For relatively simple situations such as Fig. 4.12a, where the phase is a monotonically decreasing function of $\omega$, the Nyquist test reduces to the simple statement that for stability $|T(j\omega)|$ *must be less than one when* $\angle T(j\omega)$ *is* $\pm 180°$. Amplifiers with more complicated phase behavior, such as that indicated in Fig. 4.13*, must be treated more carefully. The gain versus phase plot in this figure in the vicinity of the $-180°$ point is characteristic of a so-called *marginally stable amplifier*. Even though this type of amplifier is only of minor practical importance, it nonetheless raises some important issues of amplifier stability which must be disposed of with considerable care.

The problem with this marginally stable amplifier is that it has large gain when the phase equals $-180°$, as can be seen from the Bode plot in Fig. 4.13b. At first glance this might indicate instability. However, on the basis of the Nyquist test, the amplifier is

---

* Because the magnitude versus phase plot is always symmetrical about the axis of zero phase, it is customary to show the plot only for positive $\omega$.

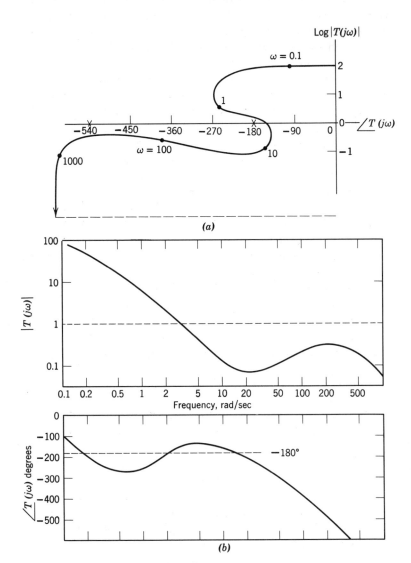

**Fig. 4.13.** $T(j\omega)$ for a marginally stable amplifier. (a) Log magnitude versus phase for $T(j\omega)$. (b) Magnitude and phase of $T(j\omega)$ versus $\omega$.

*stable*. (See Problem P4.6.) In spite of this fact, amplifiers of this type are very difficult to implement successfully, primarily because of nonlinear and other practical aspects of the amplifier performance which we have disregarded previously. There are two points of view that illustrate the problem.

In a marginally stable amplifier, if for some reason $|T(j\omega)|$ decreases while $\angle T(j\omega)$ remains unchanged, the amplifier becomes unstable. For example, the simple transient of turning on the amplifier may cause the gain to increase gradually until quiescent conditions are reached. However, the amplifier may start to oscillate before it ever reaches the stable operating point determined on the basis of the final values of the parameters.

A second point of view is to examine the root-locus plot for increasing $T(0)$. For a marginally stable amplifier, the roots will start in the left half-plane, then move into the right half-plane, and then move back to the left half-plane. A transient or overload condition may reduce the effective value of $T(0)$, thereby initiating oscillations which may persist indefinitely.

We conclude that the marginally stable amplifier is not easy to implement in practice. Thus it is reasonable to extend the above-mentioned simple form of the stability criterion to almost all feedback designs. That is, in the vast majority of feedback amplifier designs, we require that $|T(j\omega)|$ be less than one when $\angle T(j\omega)$ is $\pm 180°$.

Another feature of the log $|T(j\omega)|$ versus $\angle T(j\omega)$ plot is shown in Fig. 4.13a. According to the Nyquist Criterion, all of the points $|T| = 1$, $\angle T = \pm(2n + 1)180°$ represent possible poles of $A(s)$. Thus, in this example, the locus approaches quite close to the point $|T| = 1$, $\angle T = \pm 540°$, for $\omega \cong 500$ rad/sec. We can therefore conclude that $A(s)$ has one of its pairs of poles in the left half-plane in the vicinity of $\omega = 500$ rad/sec. In fact, we see from the locus that if the loop gain $T(0)$ for this particular amplifier were increased, it is this point which first crosses into the right half-plane (i.e., inside the contour). Thus the amplifier will begin to oscillate at a frequency of about 500 rad/sec.

### 4.3.2 *Criteria for Acceptable Transient Response*

In the root locus discussion we established the somewhat arbitrary criterion that for acceptable transient response, complex

## Sec. 4.3 Design Based on j-Axis Response

poles should not lie above the 45° radial lines or, alternatively, that the $Q$ of the complex pair should not be greater than 0.707. The corresponding criteria on the steady-state response $T(j\omega)$ are equally arbitrary, and subject to similar limitations. Two common criteria are *gain and phase margins*, and *maximum amount of peaking of the steady-state response*.

The nearness of approach of the $T(j\omega)$ contour to the $T = -1$ points is, in some sense, a measure of the degree of stability of the feedback amplifier. The parameters "gain margin" and "phase margin," shown on both the log magnitude versus phase plot and the corresponding Bode plot in Fig. 4.14, are commonly used criteria to describe the nearness of approach to these points. The gain margin is defined as $1/|T(j\omega)|$ when $\angle T(j\omega)$ is $\pm 180°$, so in this figure it is two, or 6 db. The phase margin is the angle from the $|T(j\omega)| = 1$ point to $-180°$, which for this locus is 45°.

Typical margins for practical amplifiers are 12 db and 60°. To some extent, the size of these margins is indicative of the sensitivity of the amplifier stability to parameter variations and, in addition, the margins give a rough guide to the amount of damped oscillation or ringing present in the step response. It is possible, however, to design circuits which have large gain and phase margins, but are also very sensitive to parameter variations. Thus gain and phase margins must be used with discretion.

A second criterion often used to ensure acceptable transient response is that the closed-loop amplifier response $|A(j\omega)|$ have only a small peak in it. From Fig. 4.2, it is clear that for a simple two-pole function, this criterion is identical to a $Q$ specification on the pole locations. For more complicated functions, the relationship becomes correspondingly more complicated. Nonetheless an often-used specification to ensure satisfactory transient response is that the peak of the amplitude-versus-frequency response function be less than a factor of 1.1 above the mid-frequency response. A factor such as this obviously depends to a great extent on the intended application. For example, in servo systems, factors as large as 1.3 or 1.5 are sometimes used.

### 4.3.3 Types of Plots

In the course of the preceding discussion we have shown examples of the two most common plots used in feedback amplifier analysis

**132  Stability and Frequency Response of Feedback Amplifiers**

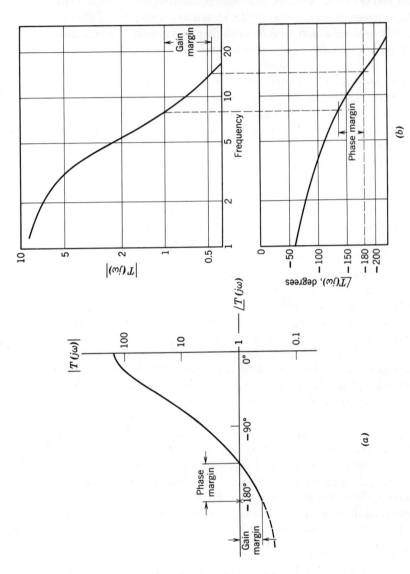

**Fig. 4.14.** Gain and Phase Margins.

and design: the Bode plot of the magnitude and phase of the loop transmission $T(j\omega)$ versus $\omega$; and the plot of the log magnitude of $T(j\omega)$ versus its phase, with frequency as a parameter. Both of these plots are convenient to use because they have logarithmic amplitude scales. The Bode plot has a logarithmic frequency scale as well, thus permitting rapid construction of the amplitude and phase curves, as we saw in Chapters 1 and 2. An important advantage of the log magnitude versus phase plot is that the magnitude and phase of the closed-loop gain, $A(j\omega)$, can be read off the plot with the aid of an overlay called a Nichols chart, shown in simplified form in Fig. 4.15. The reader is referred to any standard text on feedback for more information.

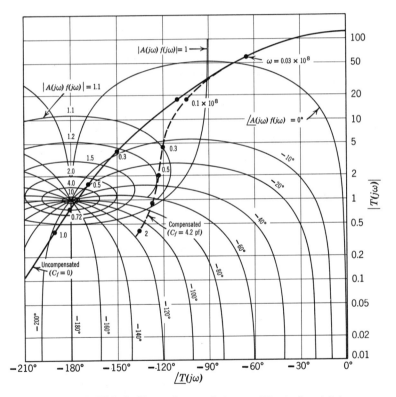

Fig. 4.15. Nichols Chart. Contour is for amplifier in Sec. 4.3.4.

### 4.3.4 Four-Stage Feedback Amplifier Example

A four-stage amplifier example will now be discussed as an illustration of the use of Bode plots and log magnitude-phase plots in feedback amplifier analysis. We will take as the basis of this example the four-stage amplifier with transadmittance feedback, shown in Fig. 4.16.

In the design, dc feedback has been used to stabilize the operating points. This method of biasing avoids phase shift due to coupling and bypass capacitors and thereby prevents low-frequency

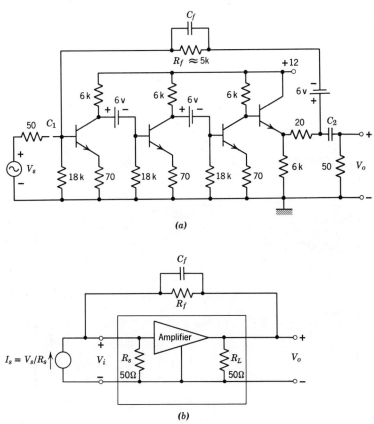

**Fig. 4.16.** Four-stage feedback amplifier example. (Resistor values in ohms unless otherwise marked.)

## Sec. 4.3 Design Based on j-Axis Response

oscillations when feedback is applied. Note that each transistor is automatically biased at $V_{CE} \approx 6$ volts, $I_C \approx 1$ milliampere. In actual practice, the 6-volt batteries would be replaced by 6-volt Zener diodes or their equivalent. Capacitors $C_1$ and $C_2$ must be chosen with care to prevent a peaked low-frequency response when the feedback is applied. The analysis and design procedure for good low-frequency performance with feedback is identical in principal to the high-frequency procedure, so again we shall discuss only the high-frequency problem.

We assume that the gain of the basic amplifier at high frequencies can be approximated by the expression

$$a(s) = \left.\frac{V_o}{I_s}\right|_{y_{rf}=0} = \frac{-(125)(100)(50)}{(1 + 59.6s)(1 + 5s)(1 + 0.42s)}$$

$$= \frac{-(125)(100)(50)}{1 + 65s + 325s^2 + 125s^3} \text{ ohms} \quad (4.31)$$

[s in units of $(\text{sec})^{-1} \times 10^8$]

These numbers are based on a circuit design using typical high-frequency transistors. The source and load resistors in the circuit are so low that there are only three dominant poles.

The voltage gain $V_o/V_s$ at midband without feedback is

$$\frac{V_o}{V_s} = \frac{V_o}{I_s R_s} = -125 \times 100$$

If we design for a closed-loop gain magnitude of 100, we can sacrifice a voltage gain of 125 by feedback. Thus we can expect to achieve a substantial improvement in desensitivity of the amplifier to parameter changes, and a corresponding decrease in distortion.

To achieve a mid-band voltage gain of $-100$, we require a transimpedance of

$$A(0) = \left.\frac{V_o}{I_s}\right|_{\text{mid-band}} = \frac{a(0)}{1 + a(0)f(0)} = -(100)(50) \text{ ohms}$$

$$\therefore 1 + T(0) = 1 + (125)(100)(50)f(0) = 125$$

$$-f(0) = G_f = \frac{124}{(125)(100)(50)} = 0.198 \text{ mmho}$$

$$\therefore R_f = 5.04 \text{ k}$$

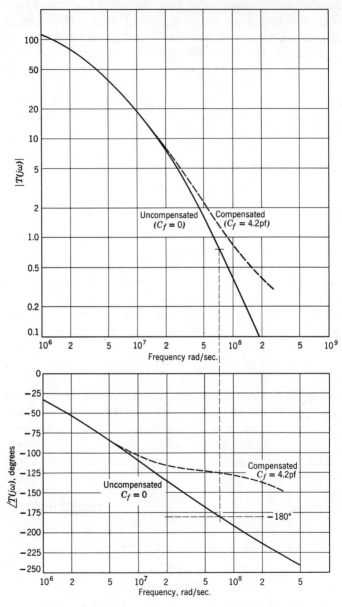

**Fig. 4.17** (a). Bode plot of $T(j\omega)$ for feedback amplifier in Fig. 4.16.

### Sec. 4.3 Design Based on j-Axis Response

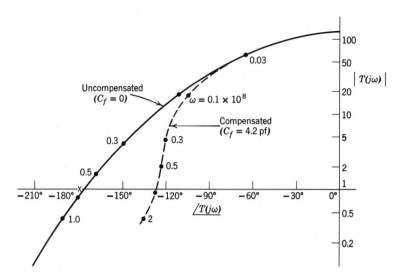

**Fig. 4.17** (b). Log magnitude versus phase for $T(j\omega)$.

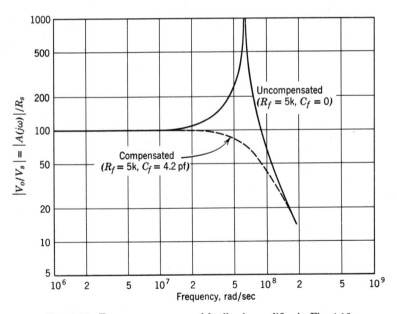

**Fig. 4.18.** Frequency response of feedback amplifier in Fig. 4.16.

Since we are striving for an over-all gain of 100, we could have estimated that $R_f \cong 100R_s = 5$ k, but, since $a$ is finite, $R_f$ must be slightly greater than 5 k.

For $C_f = 0$ and $R_f \cong 5$ k in Fig. 4.16, we find from Eq. 4.31:

$$T(s) = \frac{(125)(100)(50)(0.198 \times 10^{-3})}{(1 + 59.6s)(1 + 5s)(1 + 0.42s)} \quad (4.32)$$
[$s$ in units of (sec)$^{-1} \times 10^8$]

A Bode plot of this equation is shown in Fig. 4.17a. Note that the gain margin and phase margin are very small, so we can expect a very peaked closed-loop response. The same conclusion can also be reached from the corresponding log magnitude versus phase plot of $T(j\omega)$, Fig. 4.17b, by noting that the $T(j\omega)$ contour passes very close to the $-180°$ point.

Figure 4.18 shows the closed-loop response $|A(j\omega)|/R_s = |V_o/V_s|$ for the feedback amplifier, obtained either by calculating $A(j\omega)$ from Eqs. 4.32, 4.31, and 4.12, or by using the Nichols chart to find $|A(j\omega)|$ and $\angle A(j\omega)$ from $T(j\omega)$. To facilitate this latter computation, the $T(j\omega)$ contour in Fig. 4.17b has been added to the Nichols chart in Fig. 4.15. Note that the values read off the overlay correspond to $|A(j\omega)f(j\omega)|$ and thus must be multiplied by $R_f/R_s = 5.04 \times 10^3/50$ to get $|V_o/V_s|$. As expected, we find that $|V_o/V_s|$ has a completely unacceptable peak.

If we add in three more denominator terms in Eq. 4.32 corresponding to the three nondominant poles of the high frequency response, the only change which would occur in Fig. 4.17 would be a faster fall-off in phase at high frequency. The nondominant poles are usually so much higher in frequency than the dominant poles (see, for example, the amplifier calculations in Chapters 1 and 2) that they will not make a significant change in the amplitude response over the frequency range of interest, i.e., the vicinity of the 180° point. We conclude, therefore, that because of this added phase shift, the amplifier as it now stands will surely oscillate.

### 4.3.5 Compensation

To reduce the amount of peaking in the response, suppose we add a capacitor $C_f$ in parallel with $R_f$ in the feedback network. The reason for such an addition has already been discussed in Sec. 4.2.4. The loop gain now becomes:

$$T(s) = \frac{(125)(100)(50)(0.198 \times 10^{-3} + sC_f)}{(1 + 59.6s)(1 + 5s)(1 + 0.42s)} \quad (4.33)$$
[s in units of (sec)$^{-1} \times 10^8$]

We can make a fairly good estimate of the $R_f C_f$ product by recalling that $R_f$ and $C_f$ create a zero in the feedback circuit. We would like to place this zero at, roughly, $s = -\omega_0$ where $\omega_0$ is the frequency at which the response peaks. Thus a reasonable compensation procedure is to experimentally measure the peaking frequency and use an adjustable capacitor with $R_f C_f$ slightly larger than $\omega_0^{-1}$. We can then vary this capacitor experimentally until the desired step response is achieved.

On this basis, because $\omega_0$ here is $6.2 \times 10^7$ rad/sec, we choose

$$R_f C_f \cong 16 \text{ nanosec}$$

However, a more detailed calculation, which minimizes the $Q$ of the complex pole pair, indicates that a better value is

$$R_f C_f = 21 \text{ nanosec}$$

$$C_f \cong \frac{21 \times 10^9}{5 \times 10^3} = 4.2 \text{ pf}$$

A Bode plot of $T(j\omega)$ for this case has been added to Fig. 4.17a. This new plot can be obtained quite easily from the plot for $C_f = 0$ by noting that we can obtain Eq. 4.33 by multiplying Eq. 4.32 by

$$H(s) = \frac{0.198 \times 10^{-3} + sC_f}{0.198 \times 10^{-3}}$$

Hence, starting with the plots of $T(j\omega)$ from Eq. 4.32, we add logarithmically $|H(j\omega)|$ to $|T(j\omega)|$, and add $\angle H(j\omega)$ to $\angle T(j\omega)$. The same simple construction can also be used to find the new $T(j\omega)$ locus in Fig. 4.17b (and Fig. 4.15). It is clear from either plot that a substantial improvement in gain and phase margins has been realized.

The plot of closed-loop gain $V_o/V_s$ for this compensated case has been added to Fig. 4.18 for ready comparison with the uncompensated case. Again this plot can be derived by means of the Nichols chart, Fig. 4.15, or by finding the poles of $A(s)$:

$$A(s) = \frac{-(100)(50)}{1 + 2.6s + 2.6s^2 + s^3}$$
[$s$ in units of (sec$^{-1}$) $\times 10^8$]

The poles of this function are located at $s = -1 \times 10^8$, $s = (-0.8 \pm j\,0.6) \times 10^8$.

## PROBLEMS

**P4.1** If we use negative feedback (i.e., $g_n$ in Fig. 4.19) on the two-stage amplifier in Problem P2.1 in order to reduce the gain to 10, what is the $Q$ of the resulting complex pole pair? Can negative feedback and shunt loading be advantageously employed together in this feedback design?

**P4.2** The block diagram of a feedback amplifier is shown in Fig. 4.20. Assume that

$$a_v = \frac{10^5}{(s+2)(s+10)(s+20)}$$
($s$ in units of microseconds$^{-1}$)

$$V_i = V_1 - V_f$$

(a) If $V_f/V_o$ equals $f_0$ (a constant), how large can $f_0$ be made without producing instability?

(b) If $f = f_0 \left[ \dfrac{s+20}{20} \right]$, what value of $f_0$ will cause the amplifier to have a pair of poles on the 45° radials? Calculate the location of *all* the poles of $V_o/V_1$ under this condition.

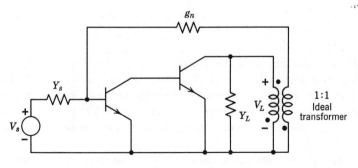

**Fig. 4.19.**

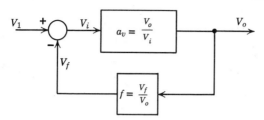

Fig. 4.20.

**P4.3** (a) The nominal low-frequency open-loop gain of the amplifier in Fig. 4.21 ($2 \times 10^3$) is subject to $\pm 10\%$ variations. Choose $f_0$ so that the variations in the closed loop gain ($V_o/V_1$) caused by the $\pm 10\%$ variations in $a$ are reduced to $\pm 0.1\%$ *at low frequencies*.

(b) Show whether or not the feedback amplifier is stable for $f_0$ equal to $\frac{1}{15}$. In the gain expression, $s$ is in units of microseconds$^{-1}$.

**P4.4** A three-stage common-emitter amplifier can be represented as shown in Fig. 4.22. Under open-loop conditions, the gain function of the amplifier is

$$a_z = \left.\frac{V_o}{I_s}\right|_{\text{open loop}} = \frac{-4 \times 10^4}{(1 + 10s)(1 + 20s)(1 + 40s)}$$
($s$ in units of microseconds $^{-1}$)

(a) If conditions are such that the closed-loop gain function of the amplifier may be written as

$$A_z = \frac{a_z}{1 + a_z f_y}$$

find $f_y$ in terms of the circuit parameters.

(b) If $Y_f$ is a conductance $G_f$, what is the maximum value of loop transmission for which the feedback amplifier is stable?

(c) Choose $G_f$ so that the feedback amplifier will have a complex pole pair on the 45° radials. Estimate the mid-band gain and the bandwidth of the amplifier with this value of $G_f$.

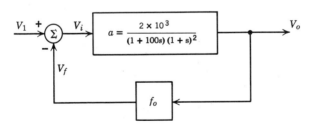

Fig. 4.21.

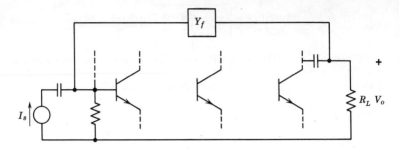

Fig. 4.22.

(d) If $Y_f$ is a parallel $GC$ network, what is a reasonable choice of $\tau_f = C/G$ for this feedback amplifier? It is desired to obtain the greatest bandwidth consistent with stability.

(e) With $\tau_f$ set at the value chosen in (d), set $G$ so that the feedback amplifier will have poles on the 45° radials. Estimate the resulting mid-band gain and bandwidth of the amplifier with these values of $\tau_f$ and $G$.

**P4.5** To illustrate the difference between Fig. 4.12 and the more conventional Nyquist test, draw the polar plot of $T(j\omega)$ to form a true Nyquist plot, obtaining values from Fig. 4.12. Indicate the region of the polar plot which corresponds to the right half-plane. Is the amplifier stable?

**P4.6** Repeat Problem P4.5 for the marginally stable amplifier in Fig. 4.13.

**P4.7** Find $I_o/V_s$ for the common-emitter amplifier with an $RC$ circuit in the emitter lead, as shown in Fig. 4.23. To illustrate what happens when there is a pole in the feedback network with this type of feedback:

(a) Carry out the analysis from the feedback point of view (in this case, transimpedance feedback; see Fig. 3.13a).

(b) Analyze the network by direct calculation (without making any feedback approximations) and compare the results. Comment on the validity of the feedback analysis in various regions of the com-

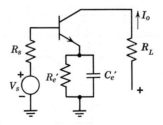

Fig. 4.23.

plex plane. In particular, does the feedback analysis give useful results for $s = j\omega$?

Problems P4.8, P4.9, and P4.10 deal with compensation techniques in transistor feedback amplifiers. They deal with modification of the gain-phase characterization, improved linearity, wider bandwidth, and control of gain without causing the feedback system to be unstable. They do *not* deal with the important question of how the amplifier circuit should be modified to accomplish the desired changes.

For each of the problems, assume that the basic amplifier has a transfer function given by

$$a(s) = \frac{V_o}{V_i} = \frac{7 \times 10^3}{(1 + 5s)(1 + 50s)(1 + 80s)}$$
($s$ in units of nanoseconds$^{-1}$)

In each problem, the general design goal is to achieve a loop gain at low frequencies which is as large as possible consistent with the requirements that the gain margin shall be at least 10 db and the phase margin shall be at least 30°.

**P4.8** Consider frequency-insensitive feedback, i.e., $f = f_0$ in Fig. 4.20 and no compensation. Determine $f_0$ and the approximate bandwidth of the amplifier with feedback by using the *design criterion* specified above. Compute and plot the over-all gain with feedback, $A = V_o/V_1$. Plot $|A|$ and $\angle A$ versus log $\omega$. Is the relative stability of the resulting amplifier satisfactory?

**P4.9** Consider lead compensation in the feedback path; that is, $f = f_0(1 + \tau s)$. Choose $\tau$ and $f_0$, and determine the low-frequency loop gain and the approximate bandwidth using the design criterion specified in the problem statement. Compute and plot the overall gain with feedback, $A = V_o/V_1$. Plot $|A|$ and $\angle A$ versus log $\omega$. Is the relative stability of the resulting amplifier satisfactory?

**P4.10** Consider lead compensation in the forward path. That is, assume that the forward transmission can be modified as shown in Fig. 4.24. Use frequency-insensitive feedback ($f = f_0$) and choose $\alpha$, $T$, and $f_0$ by using the design criterion given above. Determine the low-frequency loop gain and the approximate bandwidth of the feedback amplifier. Compute and plot the over-all gain with feedback, $A = V_o/V_1$. Plot

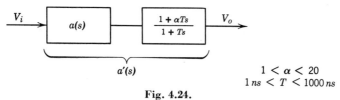

Fig. 4.24.

$|A|$ and $\angle A$ versus log $\omega$. Is the relative stability of the resulting amplifier satisfactory?

**P4.11** This problem is a continuation of Problem P3.5.
(a) If $R_f$ in Fig. 3.22b (page 102) is replaced by a capacitor of value $C_f$, the feedback amplifier becomes an integrator. For what range of frequencies is this a good integrator (i.e., $V_o \approx (1/R_i C_f) \int V_i dt$)? Assume that for a good integrator $88° < \angle V_o/V_i < 92°$ and that $\angle a = 0°$.
(b) Usually an operational amplifier is compensated so as to have $-90° < \angle a < 90°$ for $|a| > 1$. This $90°$ phase margin is necessary to insure stability for the most common feedback connections and it implies that for $|a| > 1$ the open-loop gain is approximately of the form $a = a_0(1 + s\tau)^{-1}$. Considering the effect of $\tau$, how is the answer to (a) modified.
(c) The compensation described in (b) is usually accomplished by a series $RC$ circuit between collectors of the input transistors (see Fig. 6.24, page 222) and a simplified model is shown in Fig. 4.25. Assume, for simplicity of analysis, that $y_m$ in Fig. 4.25 has a second-order pole at $s = -\tau_o^{-1}$ so that the open-loop gain is

$$a = \frac{a_0(1 + s\tau_z)}{(1 + s\tau_p)(1 + s\tau_o)^2}$$

where $\tau_z$ and $\tau_p$ are zero and pole time constants caused by $R$, $r_3$, and $C$. If $a_0$, $r_3$, and $\tau_o$ are specified, what values of $\tau_p$ and $\tau_z$ would you use to insure that the $90°$ phase margin is satisfied, and also that the open-loop bandwidth is maximum? Assume that $a_0 = 10^4$, $r_3 = 10^4$ ohms, and $\tau_o = 10^{-7}$ sec, and calculate $R$ and $C$.

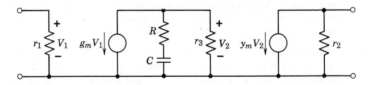

**Fig. 4.25.** Simplified high-frequency model for an operational amplifier.

*Part Two*

*Selected Topics*

# 5
# Broadband Amplifiers

## 5.0 INTRODUCTION

As an application of the methods of bandwidth calculation discussed in Chapter 1, let us study various ways of broadbanding (increasing the bandwidth of) a multistage amplifier. This problem obviously has many similarities to the single-stage problem discussed in ECP, Chapter 6, but there is one very significant difference. Any change in the load resistance of one stage of a multistage amplifier (other than the output stage) produces at the same time a change of the source resistance driving the succeeding stage; so, in a multistage amplifier, the source and load resistances of the interior stages are not independent. Consequently, the interdependence of gain and bandwidth of the amplifier, found even in an isolated single stage, is now more complicated.

To simplify the discussion, we will consider only the case where the interior stages of the multistage amplifier are identical. Under these conditions, we can define a "typical interior stage." If the typical interior stage is analyzed, the results may be used to predict the performance of the cascade. In this chapter, we determine the gain and bandwidth capabilities of this typical interior stage as we apply such broadbanding techniques as shunt resistive loading and feedback via an unbypassed emitter resistor.

## 5.1 TYPICAL INTERIOR STAGE OF A CASCADED AMPLIFIER

A portion of a cascaded transistor amplifier is shown in Fig. 5.1a. The circuit model can be represented as in Fig. 5.1b. We have assumed that the coupling capacitors $C_1$ and the bypass capacitors $C_e$ can be considered short circuits over the frequency range of interest. Also, we have omitted the elements $r_o$ and $r_\mu$ in the hybrid-$\pi$ model because the gains per stage available in the iterative amplifier designs discussed here will always be small enough to permit this approximation. On this basis, we draw the approximate circuit model of Fig. 5.2 based on the $C_t$ approximation (Sec.

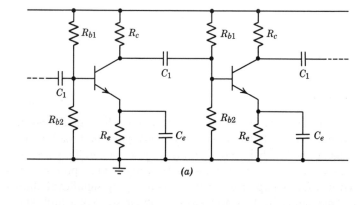

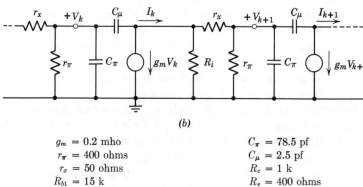

$g_m = 0.2$ mho $\qquad C_\pi = 78.5$ pf
$r_\pi = 400$ ohms $\qquad C_\mu = 2.5$ pf
$r_x = 50$ ohms $\qquad R_c = 1$ k
$R_{b1} = 15$ k $\qquad R_e = 400$ ohms
$R_{b2} = 4.7$ k $\qquad R_i \equiv R_c||R_{b1}||R_{b2} = 780$ ohms

**Fig. 5.1** Multistage video amplifier. (a) Schematic of two stages. (b) Circuit model of two stages.

### Sec. 5.1 Typical Interior Stage of a Cascaded Amplifier 149

1.4). (Remember that this model can be used to calculate forward gain and input impedance for $\omega < 3\omega_h$, *but not reverse gain or output impedance*.)

We can now designate any one repeating section of the circuit model in Fig. 5.2 as a *typical interior stage*. For this typical stage, we define from Fig. 5.2 a voltage gain as:

$$A_v = \frac{V_{k+1}}{V_k} \tag{5.1}$$

Alternatively, we can define a current gain:

$$A_i = \frac{I_{k+1}}{I_k} \tag{5.2}$$

Notice that these gain functions are not taken between the same two points in the network. However, for the definitions above, and if the $g_m$ of the transistors are equal, the two gain functions are equal:

$$A_i = \frac{-g_m V_{k+1}}{-g_m V_k} = A_v \tag{5.3}$$

For a cascade of stages, the over-all gain function is the product of these individual gain functions for the interior stages and appropriately defined gain functions for the input and the output stages of the cascade.

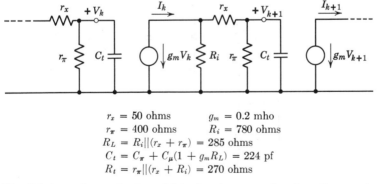

$r_x = 50$ ohms  $\quad g_m = 0.2$ mho
$r_\pi = 400$ ohms  $\quad R_i = 780$ ohms
$R_L = R_i \| (r_x + r_\pi) = 285$ ohms
$C_t = C_\pi + C_\mu(1 + g_m R_L) = 224$ pf
$R_t = r_\pi \| (r_x + R_i) = 270$ ohms

**Fig. 5.2** Approximate circuit model for interior stages, based on $C_t$ approximation. (Good only for forward gain and input impedance calculations.)

## 5.2 RESISTIVE BROADBANDING

For a given transistor type in the cascade amplifier design shown in Fig. 5.1, there are many ways in which we can adjust the bandwidth. For example, we can change the value of $R_c$, or change the operating point of the transistors, or introduce additional circuit elements such as an emitter resistor $R_e'$.

In this section, we study the effect on the gain and bandwidth brought about by changing the load resistor $R_c$. We will assume that the dc operating point of the transistor is held constant by suitably adjusting the value of $R_d$ in Fig. 5.3 to compensate in the dc circuit for the change in $R_c$. The ac circuit models of Figs. 5.1b and 5.2 are, of course, still applicable.

The mid-band current-gain expression for the typical interior stage is, from Fig. 5.2:

$$A_{i0} = \frac{-g_m r_\pi R_L}{r_x + r_\pi} \tag{5.4}$$

where $R_L$ is the resistive load on the stage, defined as:

$$R_L = R_i || (r_x + r_\pi) \tag{5.5}$$

The open-circuit time constant of the interior stage is:

$$C_t R_t = [C_\pi + C_\mu(1 + g_m R_L)]R_t \tag{5.6}$$

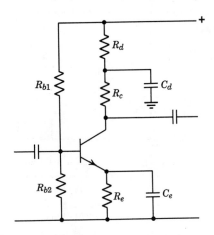

**Fig. 5.3.** Typical stage, illustrating method of changing ac load $R_c$ while maintaining same collector current.

where resistor $R_t$ is the resistance facing $C_\pi$ in Fig. 5.1b, calculated with all other capacitors open. Alternatively, it is the Thévenin equivalent resistance faced by the capacitor $C_t$ in Fig. 5.2. In either case,

$$R_t = r_\pi || (r_x + R_i) \tag{5.7}$$

If a number of these typical stages were cascaded, then according to our approximate analysis, Sec. 1.4, the bandwidth of $n$ interior stages the amplifier would be, from Eqs. 1.42 and 1.75,

$$\omega_h \cong \frac{1}{\sum R_t C_t} = \frac{1}{n R_t C_t} \tag{5.8}$$

if we ignore any drop-off in amplitude at low frequencies caused by the coupling and bypass capacitors. For convenience we shall refer to a "bandwidth" of the interior stage as:

$$n\omega_h \equiv \frac{1}{R_t C_t} = \frac{1}{R_t [C_\pi + C_\mu (1 + g_m R_L)]} \tag{5.9}$$

Then, in accordance with Eq. 5.8, we find the approximate bandwidth of the over-all amplifier by dividing this "per-stage" bandwidth by $n$. Equation 5.9 thus provides a meaningful indication of over-all amplifier performance, *but we must recognize that no one stage in the amplifier will actually have a bandwidth of this amount.* It is clear from Eqs. 5.4 and 5.9 that the gain and bandwidth of the stage are interrelated, and that this interdependence is not simple, because $R_i$ appears in both $R_L$ and $R_t$. Let us examine these relationships in considerably more detail.

### 5.2.1 *Gain and Bandwidth Performance of the Typical Interior Stage*

To provide a vehicle for the discussion, let us examine the gain and bandwidth of the typical interior stage shown in Fig. 5.2. For convenience we have used the same element values given in Sec. 1.0. These values are given in Figs. 5.1 and 5.2.

Using the numbers from Fig. 5.2, the mid-band current gain of the typical interior stage is, from Eq. 5.4,

$$A_{i0} = \frac{-g_m r_\pi R_L}{r_x + r_\pi} = \frac{-0.2 \times 400 \times 285}{450} \cong -50$$

## 152 Broadband Amplifiers

The approximate "bandwidth" of the interior stage is, from Eq. 5.9, and Fig. 5.2,

$$n\omega_h = \frac{1}{R_t C_t} = \frac{1}{(270)(224 \times 10^{-12})} = 1.65 \times 10^7 \text{ rad/sec}$$

An approximation to the response of the typical interior stage is shown in Fig. 5.4. Here we have plotted $|I_{k+1}/I_k|$ versus $\omega$, calculated on the basis of the $C_t$ approximation, Fig. 5.2. (This calculation is analogous to that made in Sec. 1.5 for $|A_v|$ of a complete amplifier.) The response beyond the 0.707 point is shown as a dashed line, indicating that the analysis is not accurate in this region.

If the gain and bandwidth shown in Fig. 5.4 are not satisfactory, then one way of changing the bandwidth is to change $R_i$ in each stage by changing $R_c$. (It is conceptually simpler to think in terms of varying $R_i$, because the gain and time-constant equations are written in terms of this parameter. Hence, from now on in this section we will talk in terms of varying $R_i$, realizing that in actual fact $R_i$ would be varied by varying $R_c$.) The plots of $|A_i|$ versus $\omega$ for three additional values of $R_i$, namely, 10 $k$, 100 ohms, and 10 ohms, are also shown in Fig. 5.4 to illustrate the effect of $R_i$ on amplifier performance. As expected, gain can be exchanged for bandwidth by lowering $R_i$.

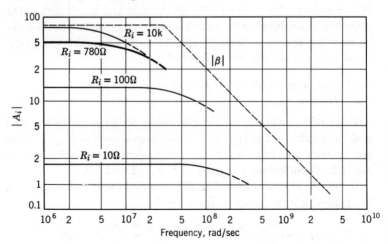

**Fig. 5.4.** Frequency response of typical interior stage for four values of $R_i$.

There are fairly definite limits, however, on the values of gain and bandwidth that can be achieved by varying only $R_i$. For example, it can be shown that for this circuit the $|A_i|$ curves in Fig. 5.4 will always fall below the $|\beta|$ versus $\omega$ curve for the transistor, regardless of the value of $R_i$. Said in another way, for this single interior stage, the gain and per-stage bandwidth coordinates, Eqs. 5.4 and 5.9, always define a point below the $|\beta|$ versus $\omega$ curve for the transistor, regardless of the value of $R_i$. Thus the asymptotes of $|\beta|$ provide a useful upper bound for the plots of $|A_i|$, and for this reason are included in Fig. 5.4 (see Problem P5.5).

The relationship between gain and bandwidth of the interior stage can be seen more directly if we plot gain versus per-stage bandwidth $n\omega_h$ on a log-log scale as shown in Fig. 5.5. Several values of $R_i$ are shown on the curve. This plot is in effect *the locus of 0.707 points for changing $R_i$*. It is obviously no longer a frequency-response plot, but in spite of this, it is still helpful to include the $|\beta|$ asymptotes on the same graph, because the $|\beta|$ curve represents an upper bound on the locus of 0.707 points.

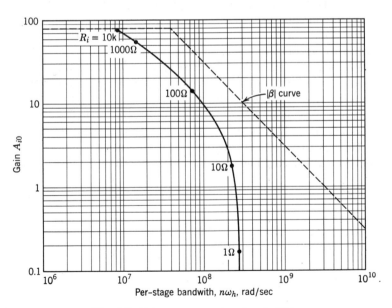

**Fig. 5.5.** Gain versus per-stage bandwidth.

We can see from Eqs. 5.4, 5.5, 5.7, and 5.9 that two separate effects determine the shape of the gain-versus-bandwidth curve. For large gain and small bandwidth (large $R_i$), the effect of $C_\mu$ becomes quite pronounced. On the other hand, for large bandwidths, and hence low gains (small $R_i$), $r_x$ and $C_\pi$ begin to dominate. For a given specified bandwidth, the curve indicates how much gain can be achieved in the typical interior stage at this particular operating point.

### 5.2.2 *Gain and Bandwidth Performance as a Function of Operating Point*

We saw in PEM* and ECP that the parameters in the hybrid-$\pi$ model vary with operating point. Therefore, it is not surprising that the curve of gain versus bandwidth shown in Fig. 5.5 also varies with operating point. The gain-versus-bandwidth curve in the preceding section was calculated for a silicon transistor at a collector current of 5 milliamperes. To give some idea of how the gain-bandwidth curves vary with operating point, let us compute these curves for the same transistor for collector currents of 1 and 10 milliamperes.

Plots of $\beta_0$ and $f_T$ as a function of collector current for this transistor are shown in Fig. 5.6. The variation of $\beta_0$ is less than 10 per cent for the currents of interest here, so for simplicity, we will assume it has a constant value of 80. Also for simplicity, we will consider that $r_x$ is a constant over this range of currents. If more accuracy were desired, variations with current of all of the hybrid-$\pi$ parameters could, of course, be taken into consideration.

On this basis we can calculate the values of $g_m$, $r_\pi$, and $C_\pi$ as a function of current. The results are given in Table 5.1. We can now

**TABLE 5.1**

| $I_C$ | $f_T$ | $g_m$ | $r_\pi$ | $C_\pi$ |
|---|---|---|---|---|
| 1 ma | 150 mc | 0.04 mhos | 2000 ohms | 40 pf |
| 5 | 394 | 0.2 | 400 | 78.5 |
| 10 | 375 | 0.4 | 200 | 170 |

* *Physical Electronics and Circuit Models of Transistors*, by P. E. Gray, D. DeWitt, A. R. Boothroyd, and J. F. Gibbons, hereafter referred to as PEM.

### Sec. 5.2 Resistive Broadbanding

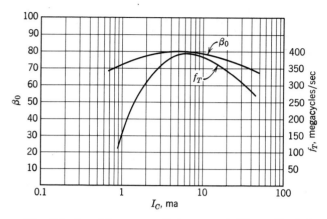

**Fig. 5.6.** $\beta_0$ and $f_T$ versus collector current, ($V_{CE} = 5$ volts).

calculate for these three new current levels the gain and per-stage bandwidth for various values of $R_i$. These calculations have been plotted in Fig. 5.7. For convenience, the plot corresponding to the 5-milliampere operating point treated in the preceding section is also included.

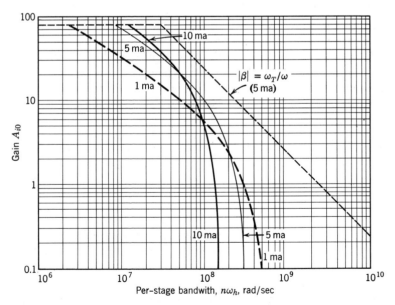

**Fig. 5.7.** Gain versus per-stage bandwidth for $I_C = 1$ ma, 5 ma, and 10 ma.

## 156  Broadband Amplifiers

From the figure, we see that no single operating current yields maximum gain for all bandwidths. However, some rough design guidelines can be laid down for operation at the high gain, narrow bandwidth end of the curve. For large $R_i$, $R_L$ approaches $r_x + r_\pi$, while $R_t$ approaches $r_\pi$ (see Eqs. 5.5 and 5.7). Thus, from Eq. 5.4, the gain of the typical interior stage approaches $g_m r_\pi$ or $\beta_0$ (which we have assumed invariant with collector current). From Eq. 5.9, we see that, under the same conditions of large $R_i$, the per-stage bandwidth can be rewritten as:

$$n\omega_h\big|_{R_i \gg r_\pi} = \frac{1}{\{C_\pi + C_\mu[1 + g_m(r_x + r_\pi)]\}r_\pi} \quad (5.10a)$$

$$= \frac{1}{\beta_0 \left(\dfrac{1}{\omega_T} + r_x C_\mu + r_\pi C_\mu\right)} \quad (5.10b)$$

Because for large $R_i$ the gain is the same regardless of collector current, maximum performance results when the collector current is chosen solely on the basis of maximum bandwidth. In Eq. 5.10b, only $\omega_T$ and $r_\pi$ vary appreciably with current. A little thought will show that, for this case, maximum bandwidth will be achieved at a current somewhat *larger* than that which yields maximum $\omega_T$.

The curves of Fig. 5.7 provide a useful if somewhat cumbersome procedure for selecting an operating point to provide maximum performance for a given bandwidth specification. If, for example, the per-stage bandwidth were to be 10 megacycles, then the figure would indicate that a 5-milliampere collector current would yield highest gain. Note, however, that for this particular bandwidth, operating at 10 milliamperes would provide almost identical performance within the accuracy of these calculations.

### 5.2.3 *Plots of Gain Versus* $(n\,\omega_h)^{-1/2}$

Clearly, it involves considerable work to plot the family of curves shown in Fig. 5.7 for each possible transistor that might be used in a circuit, in order to select a transistor and an operating point. Fortunately, the equations can be recast in a form which makes plotting much easier. It is evident from Fig. 5.7 that each of the curves for a given transistor approaches a limiting bandwidth value at low gain. This limiting value can be found from Eqs. 5.9, 5.5, and 5.7 by setting $R_i$ equal to zero:

$$n\omega_h\big|_{R_i=0} = \frac{g_x + g_\pi}{C_\pi + C_\mu} \equiv \omega_b \quad (5.11)$$

a value equal to the *transverse cut-off frequency* $\omega_b$ defined in Chapter 1. Equation 5.11 can be seen directly from Fig. 5.2, because it is clear from the figure that, regardless of how small we make $R_i$, the Thévenin equivalent resistance seen by $C_\pi + C_\mu$ cannot become less than $r_x \| r_\pi$.

We can use Eq. 5.11 to locate readily one point on the gain versus per-stage bandwidth curves. In addition, it turns out that each curve in Fig. 5.7 can be approximated by a straight line if we are willing to replot the functions as gain versus $(n\omega_h)^{-1/2}$, as shown in Fig. 5.8. The mathematical justification for this approximation, which will not be given here, involves substituting for $R_L$ and $R_t$ in Eqs. 5.4 and 5.9 from Eqs. 5.5 and 5.7, then combining the re-

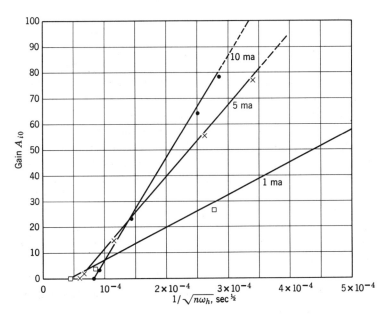

**Fig. 5.8.** Gain versus $1/\sqrt{n\omega_h}$. Straight lines are plots of Eq. 5.12. Individual points are taken from Fig. 5.7.

sulting equations to eliminate $R_i$. By approximating the resulting quadratic in $A_{i0}$ by a perfect square in such a way that the approximation is asymptotically correct for large and small values of $R_i$, we obtain the simple linear relationship

$$A_{i0} \cong \sqrt{\frac{g_m}{C_\mu}} \left( \frac{1}{\sqrt{n\omega_h}} - \frac{1}{\sqrt{\omega_b}} \right) \qquad (5.12)$$

The three lines shown in Fig. 5.8 are plots of this equation, while the individual points on the figure correspond to the plots in Fig. 5.7. The quality of the straight-line approximation in Eq. 5.12 is evident.

We can conclude that Eq. 5.12 provides a very simple way of intercomparing transistors and selecting operating currents to meet a given design requirement. To plot a curve we need only calculate the slope:

$$\text{Slope} = \frac{\Delta A_{i0}}{\Delta (1/\sqrt{n\omega_h})} = \sqrt{\frac{g_m}{C_\mu}} \qquad (5.13)$$

and the intercept:

$$X\text{-axis Intercept} = \frac{1}{\sqrt{n\omega_h}}\bigg|_{A_{i0}=0} = \frac{1}{\sqrt{\omega_b}} = \sqrt{\frac{C_\pi + C_\mu}{g_x + g_\pi}} \qquad (5.14)$$

Once these plots are made, it is obviously a simple matter to determine which of several transistors is best for a given design, and then which of several operating points is best.

In any particular design, the approximations made in deriving the above equations must always be kept in mind. In particular, if the gain-versus-bandwidth plots in a particular design call for very small values of $I_C$, especially less than 1 milliampere, then it must be borne in mind that $\beta_0$ will be decreasing when the collector current becomes this small. Thus, in many practical situations, no improvement in performance will be achieved by going to collector currents less than 1 milliampere, in spite of what our approximate analysis may indicate.

## 5.3 BROADBANDING BY ADDITION OF EMITTER RESISTANCE

We now discuss two methods by which the gain-versus-bandwidth plots for a given transistor can be modified by adding simple

## Sec. 5.3 Broadbanding by Addition of Emitter Resistance

passive elements to the basic interior stage. In this section, we analyze the effects of adding a resistor in series with the emitter lead of the transistor. In Sec. 5.4, we discuss the effects of adding an inductor to the interstage network.

It was shown in Sec. 5.2 that, for large-bandwidth, low-gain stages, $r_x$ and $C_\pi$ become the limiting factors in the bandwidth expression. Also we saw in Fig. 5.7 that operating at a lower current slightly improves performance in this region because $C_\pi$ is reduced.

If a series emitter resistor $R_e'$ is introduced in the circuit, as shown in Fig. 5.9a, we have a way of reducing the effective value of $C_\pi$ without substantially changing the effective $\omega_T$, as was pointed out in Sec. 2.4. The small-signal model for this circuit is shown in Fig. 5.9b. Using the relations developed in Sec. 2.4.1 and summarized in Fig. 2.11c, we replace the elements within the dashed contour by $r_\pi'$, $C_\pi'$ and $g_m'$, where

$$r_\pi' = \frac{r_\pi}{k_2} \tag{5.15}$$

$$C_\pi' = k_2 C_\pi \tag{5.16}$$

$$g_m' = k_2 g_m \tag{5.17}$$

$$k_2 = \frac{1}{1 + g_m R_e'} \tag{5.18}$$

The resulting circuit is shown in Fig. 5.9c.

Because we again have the original transistor configuration for the typical interior stage of the cascade, the developments and conclusions in the preceding sections concerning the gain level and bandwidth relationships remain unchanged. In particular, we can still use the $C_t$ approximation to find the bandwidth, except that we must use the modified value $C_t'$ defined as

$$C_t' = C_\pi' + C_\mu(1 + g_m' R_L) \tag{5.19}$$

Now, however, we are free to choose $R_e'$ to obtain small values of $C_\pi'$ and $g_m'$, and thus improve high-frequency performance.

### 5.3.1 Example

To illustrate the effect of $R_e'$ on the gain and bandwidth performance of the typical interior stage, we calculate the gain-versus-

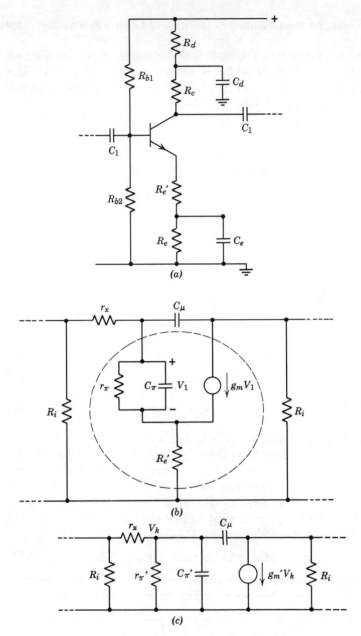

**Fig. 5.9.** Transistor with series emitter resistor.

## Sec. 5.3 Broadbanding by Addition of Emitter Resistance 161

bandwidth curve for the same transistor discussed in the preceding sections. We will operate the transistor at 5 milliamperes collector current, with 20 ohms of resistance inserted in the emitter lead. Using Eqs. 5.15 through 5.18, we find that

$$k_2 = \frac{1}{1 + (0.2)(20)} = \frac{1}{5}$$

Therefore, the new values for the hybrid-$\pi$ parameters are:

$$g_m' = 0.04 \qquad r_x = 50 \text{ ohms}$$
$$r_\pi' = 2000 \text{ ohms} \qquad C_\mu = 2.5 \text{ pf}$$
$$C_\pi' = 15.5 \text{ pf}$$

The results of the gain-versus-bandwidth calculation for this case are shown in Fig. 5.10. The curve for $I_C = 5$ ma, $R_e' = 0$ is included on the same graph for reference. Comparison with Fig. 5.7 shows that the introduction of $R_e'$ has, as a matter of fact, opened up a new area of operation with even greater bandwidth capabilities than previously obtained with 1-milliampere bias and $R_e'$ equal to zero.

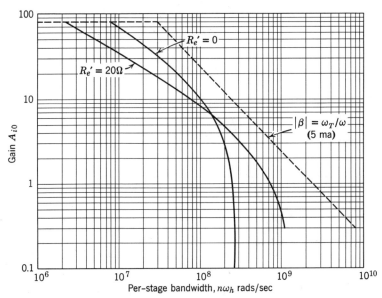

**Fig. 5.10.** Gain versus per-stage bandwidth with emitter resistor, ($I_C = 5$ ma).

### 5.3.2 Selection of $I_C$ and $R_e'$

The plots of Fig. 5.10 are useful for visualizing circuit performance, but the calculations involved in deriving a set of curves are tedious. However, we can again make good use of the approximately linear relationship between gain and $(n\omega_h)^{-1/2}$ given by Eq. 5.12. Figure 5.11 shows three such plots corresponding to three different values of emitter resistor, namely, 0, 20, and 45 ohms, all calculated for $I_C = 5$ ma from the slope and intercept values of Eqs. 5.13 and 5.14.

The data presented here are, of course, no different from those shown in Fig. 5.10. Thus the conclusions are the same as before; namely, that for very large bandwidths, introduction of a small resistor in the emitter lead does provide an important improvement in gain. It is clear from the figure, however, that increasing $R_e'$ from 20 ohms to 45 ohms for this particular transistor does not provide significant improvement except in extreme cases. Increas-

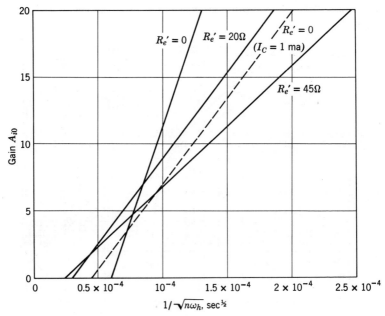

**Fig. 5.11.** Gain versus $1/\sqrt{n\omega_h}$ with emitter resistor. For solid curves, $I_C = 5$ ma; dashed curve, 1 ma.

ing $R_e'$ much beyond 45 ohms would result in a bandwidth improvement only at gains less than one, and this is of no interest at all.

Also shown in Fig. 5.11 is a dashed curve for $I_C = 1$ ma, $R_e' = 0$, because the $g_m$ for this case equals the $g_m'$ for the case with $I_C = 5$ ma, $R_e' = 20$ ohms. Hence, from Eq. 5.13, the curves have the same slope. Comparison of these two curves shows that the circuit with $I_C = 5$ ma, $R_e' = 20$ ohms is everywhere better than the circuit with $I_C = 1$ ma, $R_e' = 0$. A somewhat more general statement can be made by looking at Eqs. 5.13 and 5.14. If we compare a series of designs which have been adjusted to have the same $g_m'$ by including a suitable emitter resistor, the plots of $A_{i0}$ versus $\sqrt{n\omega_h}$ for these circuits will all have the same slope. Under these conditions, the best circuit will be the one that has the smallest intercept, as given in Eq. 5.14. Because we fixed $g_m'$, this requirement amounts to designing for the smallest value of $C_\pi$ or $C_\pi'$, which in turn means *choosing $I_C$ to give the largest value of $\omega_T$*.

Of course, we cannot *increase* the $g_m'$ by adding an emitter resistor, so better performance can always be achieved in the narrow-bandwidth, high-gain region of the curve by going to somewhat higher currents, as was discussed in Sec. 5.2.2.

These conclusions are summarized in Fig. 5.12. The curve for maximum $\omega_T$, in this example the 5-milliampere curve, has been drawn as a heavy line because it represents the transition between two different design situations for this transistor. For gains less than about 15 or values of $n\omega_h$ greater than about $8 \times 10^7$ rad/sec, this transistor should be operated with $I_C = 5$ ma and the emitter resistor chosen to give the desired performance. For gains greater than 15, or bandwidths less than $8 \times 10^7$ rad/sec, the transistor should be operated with no emitter resistor and with collector currents larger than 5 milliamperes.

## 5.4 THE SHUNT-PEAKED AMPLIFIER

### 5.4.1 *Shunt Peaking*

As a final example of broadbanding techniques, we study in this section the effect of adding an inductor in the collector circuit of each stage of the amplifier, as shown in Fig. 5.13. The circuit is often called a shunt-peaked amplifier.

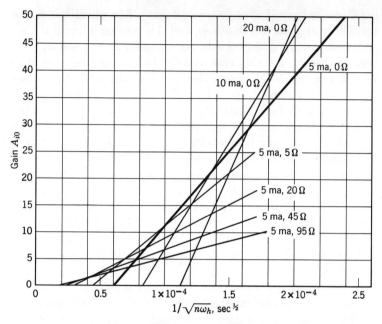

**Fig. 5.12.** Gain versus $1/\sqrt{n\omega_h}$, showing optimum choice of $I_C$ and $R_e'$.

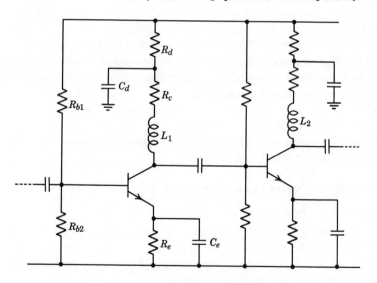

**Fig. 5.13.** Shunt-peaked stage.

### Sec. 5.4 The Shunt-Peaked Amplifier

To gain some insight into the effect of the inductor, we study the sinusoidal response of the circuit shown in Fig. 5.14. This circuit would be strictly correct *only if the entire loading on the second transistor were purely resistive*. When $L_1 = 0$ under such circumstances, the voltage $V_{k+1}$ drops as the input frequency increases because of the shunting effect of $C_t$. The 0.707 frequency is reached when the reactance of $C_t$ is equal to the resistance $R_t$ seen by $C_t$. Now with $L_1$ present, the low-frequency response is the same as for $L_1 = 0$, but as the frequency is raised, a low $Q$ resonance occurs between the inductance $L_1$ and the capacitance $C_t$. This resonance holds up or "peaks" the response of $V_{k+1}$. Plots of the frequency response of the circuit for various values of $L_1$ are shown in Fig. 5.15. As might be expected, if $L_1$ is too large, over-compensation or peaking can be obtained.

Because of problems of interaction among a number of cascaded shunt-peaked stages, amplifiers of the type shown in Fig. 5.13 are difficult to analyze and to align. In fact, it may be easier to align them experimentally than to design them by calculation. However, as a rough guide in the initial design, it is reasonable to choose each inductance so that it resonates with the $C_t$ of the following stage in the general vicinity of the 0.707 frequency desired for the peaked stage.

#### 5.4.2 *Constant-Resistance Cascade*

One way of avoiding some of the difficulties of the simple shunt-peaked amplifier brought about by the interaction between stages—difficulties of alignment and instability, for example—is to design

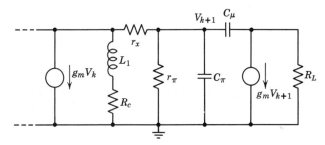

**Fig. 5.14.** Circuit model for one shunt-peaked stage, assuming $L_2 = 0$ and following stage has resistive input impedance.

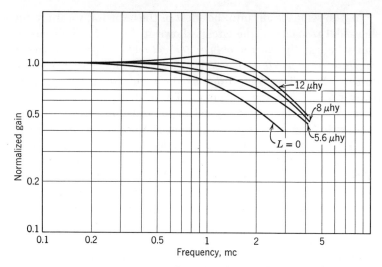

**Fig. 5.15.** Amplitude versus frequency for one shunt-peaked stage.

a multistage amplifier such that the input impedance of each stage is a constant resistance. For such a circuit, called a *constant-resistance cascade*, representing the feedback effect of $C_\mu$ on the input node by a capacitor $C_t$ is an almost exact calculation, so the behavior of the amplifier can be predicted with considerable accuracy.

We achieve a constant input resistance by adjusting the value of $L$ and $R_c$ in the input circuit for the stage, Fig. 5.16a. To find the necessary conditions for achieving a constant input resistance, we note first that a circuit of the type shown in Fig. 5.16b can be made to have constant input resistance by setting $R_1 = R_2$, to yield the same resistance at high and low frequencies, and choosing $L$ and $C$ to make the short-circuit time constants of the two branches equal.

The arrangement $r_x$, $r_\pi$, and $C_t$ in Fig. 5.16a can be replaced by a series $RC$ in parallel with a resistor without altering the admittance at all. That is, the two circuits can be made to have the same pole and the same conductance values at high and low frequencies. Thus the circuit of Fig. 5.16a can also be made to have a constant input resistance. We choose $R_c$ to yield the same input resistance at high and low frequencies, and choose $L$ to

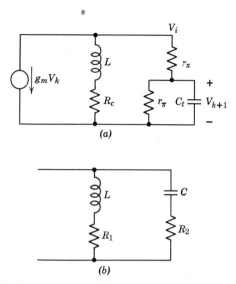

**Fig. 5.16.** Input circuit model for constant-resistance cascade.

fulfil the condition on the short-circuit time constants. (See Problem P5.4.) However, having achieved a constant-resistance cascade by this method, we no longer have a circuit element to vary for selection of gain or bandwidth. Thus to achieve other bandwidths, we must vary the transistor parameters, either by varying $I_C$ or by adding a series resistor $R_e'$ to the emitter.

The constant-resistance approach does not necessarily produce any optimum of gain or bandwidth performance. On the contrary, some sacrifice in such performance may be required to achieve the stability and alignability advantages. For this reason, true constant-resistance cascades are rarely used, but the circuit designer may well want to choose element values which approach this condition, in order to reduce interaction and simplify alignment.

## PROBLEMS

**P5.1** For the amplifier shown in Fig. 5.17, assume that the resistors $R_1$ and $R_2$ have been adjusted so that each transistor operates at $I_C = 10$ ma. At this operating point the transistor parameters are $\beta_0 = 100$, $r_x = 50$ ohms, $f_T = 155$ mc, $C_\mu = 3$ pf. Assume $R_c = 1$k.

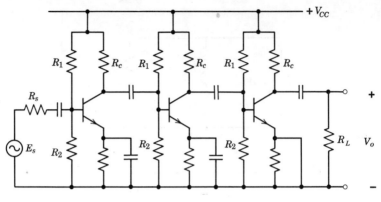

Fig. 5.17.

(a) Find the incremental model of the amplifier, assuming signal voltage drops across coupling and bypass capacitors are negligible and the biasing resistors $R_1$ and $R_2$ are much larger than $r_x$ and $r_\pi$.

(b) Calculate the gain and estimate the upper 0.707 frequency $\omega_h$. Assume that the resistor $R_L$ represents the loading effect of a broadband output stage; that is, assume that $R_L$ is always equal to the low-frequency input resistance of the identical interior stages; and $R_s$ represents the output resistance of a broadband input stage, i.e., $R_s = R_c$. Then for calculating $\omega_h$ we can consider the circuit to have three identical "interior" stages.

**P5.2** Increase the bandwidth of the amplifier in Problem P5.1 by a factor of three by decreasing the resistors $R_c$ (and $R_s$). Assume the dc operating points are unchanged. What is the over-all gain of this amplifier?

**P5.3** Assume the bandwidth of the amplifier in Problem P5.1 must be increased by a factor of 5. Redesign the amplifier (a) by decreasing $R_c$; (b) by changing the bias currents (keeping them all identical); (c) by adding identical unbypassed emitter resistors to each stage. Which way gives more gain for this amplifier? Assume $\beta_0$ and $r_x$ are constant over the range of bias currents used here, and assume $f_T$ varies with current as shown in Fig. 5.18. Omit all dc bias calculations.

**P5.4** (a) Find the relations among the parameters in Fig. 5.16b to yield a constant input resistance $R$. Then find the corresponding conditions in Fig. 5.16a on $R_c$ and $L$ in terms of $r_x$, $r_\pi$, and $C_t$.

(b) After choosing $R_c$ and $L$ in Fig. 5.16a to yield a constant input resistance, find $V_i$ in terms of $V_k$, and hence the voltage gain $A_v(s) = V_{k+1}/V_k$. From this, identify the low frequency gain per stage and the bandwidth per stage. Compare with Eqs. 5.4 and 5.9.

(c) What will be the bandwidth of an $n$-stage constant-resistance cascade? Compare with Eq. 5.8 and explain.

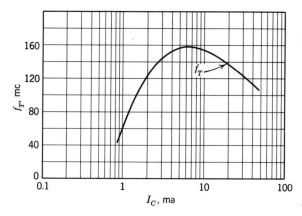

Fig. 5.18.

**P5.5** Prove that for the circuit model of the single interior stage, Fig. 5.2, the coordinates $A_{i0}$, $n\omega_h$ (i.e., Eqs. 5.4 and 5.9) for $0 < R_i < \infty$ represent a point which always lies below the $|\beta|$ versus $\omega$ curve.

# 6

# *Direct-Coupled Amplifiers*

## 6.0 INTRODUCTION AND BASIC CONSIDERATIONS

Most of the preceding chapters have dealt with ac coupled amplifiers having lower half-power frequencies greater than a few cycles per second. If it is desired to observe signals which vary very slowly, then the size and cost of the large capacitors needed for ac coupling may become impractical. Moreover, the long recovery transients associated with such capacitors often cannot be tolerated. For these reasons it is sometimes necessary to build amplifiers with pass bands which extend down to dc and to do away with all coupling and bypass capacitors.

Direct-coupled amplifiers can also be very useful even if they are intended for use only at high frequencies. For example, in very wide-band amplifiers the physical size of the coupling capacitors can produce significant problems because of their inevitable capacitance to other parts of the circuit; these capacitors can be eliminated by using direct coupling. When minimum size is desired, the use of direct coupling can reduce the number of large capacitors or inductors. Fortunately, the good long-term stability of the dc

## Sec. 6.0 Introduction and Basic Considerations

characteristics of silicon transistors has made it not only possible, but actually attractive, to use dc amplifiers even when dc amplification is not needed. This is particularly true in the case of "integrated" circuit structures wherein complex circuits are completely constructed on a single chip of silicon.

This chapter explores problems peculiar to the design and analysis of multistage amplifiers with direct coupling between stages. It does not explore all of the myriad subtleties of good direct-coupled amplifier design, but an effort has been made to exemplify the types of problems that arise, and indicate methods of analysis and design that have proven useful. It is significant that most dc amplifier problems can be handled very accurately by relatively straightforward analysis techniques. Whereas high-frequency amplifier analysis is difficult because of the complexity and inaccuracy of circuit models, dc amplifier performance can often be predicted to within a fraction of 1%. Precise analysis is often justified because it is common to use selected transistors, precision resistors, and well-regulated power supplies. It is usually *easier and more reliable* to base a design primarily on precise analysis rather than on a crude analysis followed by trial-and-error experimentation. Hence, this chapter deals not only with zero and first-order approximations, as in some of the earlier chapters, but also with second-order effects which are often the dominant factors in dc amplifier performance.

Most of this chapter will be predicated upon the use of silicon transistors. If the circuit is at all critical, the use of silicon transistors will normally lead to much better circuit performance than can be achieved with germanium transistors except, possibly, at very high frequencies. In dc amplifiers there are three important advantages of silicon transistors over germanium units: (1) $I_{CO}$ is virtually negligible, (2) a silicon oxide surface layer can be used to give very good long-term stability and predictability of the important circuit parameters, and (3) it is possible to realize large current gain at low collector currents. The disadvantages are: (1) strong temperature dependence of $h_{FE}$ ($h_{FE} \equiv I_C/I_B$ for a given $V_{CE}$) and (2) high $r_x$ at low current levels. Normally, however, the advantages are overwhelming if significant dc gain is desired.

## 6.1 TYPICAL CIRCUIT CONFIGURATIONS

### 6.1.0 *Introduction*

The choice of circuit configuration is likely to be considerably more difficult for direct-coupled amplifiers than for capacitively coupled ones. This difficulty occurs because there are more conflicting requirements on direct-coupled stages, and the proper biasing requires consideration of the whole amplifier. In ac coupled amplifiers, operating point selection and biasing circuit design are accomplished on a stage-by-stage basis. In dc amplifiers, however, signal and bias are mixed; the quiescent dc output level of one stage becomes the quiescent input level for the next stage. Thus, for example, it may be desirable to alter the quiescent output level *solely* to facilitate the coupling problem even though there is a resulting loss in gain, bandwidth, etc.

This section presents examples of multistage, direct-coupled circuit configurations and describes a suitable method of determining quiescent conditions. Symmetric or balanced circuits have specifically been avoided here because they are considered in detail in Sec. 6.3.

### 6.1.1 *Two-Transistor Circuits*

For amplifiers consisting of two cascaded stages there are only nine possible combinations of common-emitter, common-collector, and common-base stages. Six of these involve other than cascades of identical stages, and are shown in their simplest forms in Fig. 6.1. All of these combinations are useful, and many have special names. Several important characteristics of dc amplifiers can be illustrated by means of these circuits.

Note that in all of the circuits in Fig. 6.1 the operating point of the second stage is partially determined by the operating point of the first stage. Thus, in circuit (a), the collector-to-emitter voltage of the first stage is practically equal to the voltage across $R_2$. The value of $R_2$ is then uniquely determined by operating-point considerations, and we have lost a degree of design freedom which we had in ac-coupled circuits. If an additional power-supply voltage is available, then, of course, $R_2$ can be returned to this other supply and its value altered accordingly. Usually the number and values of power supply voltages are quite limited and it is mandatory to

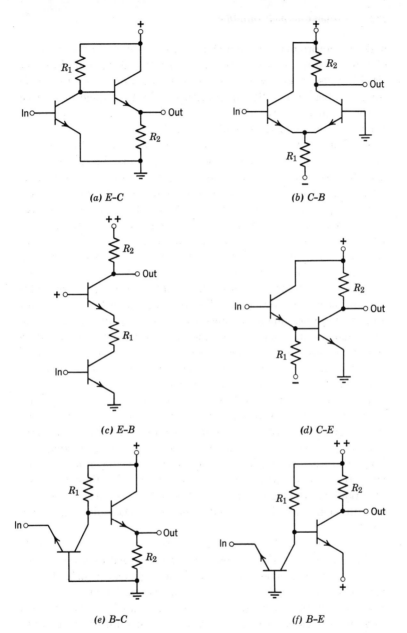

**Fig. 6.1.** Two-stage amplifiers (++, +, and − are fixed supply voltages; $E - B$ means common emitter driving common base, etc.).

make a compromise between choice of operating point and choice of load resistors. Thus, in circuit (a), if a small value of $R_2$ is desired because of bandwidth or load considerations, we must settle for a low value of collector voltage for the first stage or a large value of emitter current for the second stage. Similar considerations hold for all of the circuits in Fig. 6.1.

The bias circuitry for the input to the first transistor has been omitted in Fig. 6.1, but suitable circuitry must be employed to give desired operating conditions. The problem of devising suitable input biasing is extremely important and often difficult to solve. For example, in Fig. 6.1a we must supply both a quiescent voltage and current at the input, and usually these quiescent values must be isolated from the input signal source. Thus we might modify the input stage by returning the emitter to a small negative voltage (i.e., on the order of 0.5 to 0.7 volt if a silicon transistor is used) and adding a resistor from the plus supply to the input base. This type of biasing could, indeed, eliminate the necessity for the input source to supply quiescent power, but the resulting circuit is not very good with respect to temperature sensitivity of operating points. Ideally, the input biasing should have a precisely controlled temperature dependence which exactly compensates for temperature changes in the rest of the amplifier.

In the circuits of Fig. 6.1 the quiescent output level is not zero. Often, it is necessary to have zero output when the input is zero and, moreover, often the output and input must have the same common reference, or ground. If the direct coupling is being used only to effect a saving in coupling and bypass capacitors, then we could use a capacitor in series with the output to eliminate quiescent power in the load. For dc amplifiers, however, a series battery or equivalent must be provided to eliminate the quiescent output level.

Another class of direct coupled two-transistor circuits is exemplified in Fig. 6.2. In these circuits the two transistors are interconnected so as to behave very nearly like a single transistor and hence are sometimes called "composite" transistors. Circuit (a), sometimes called the Darlington connection, is used to achieve very high current gains. A straightforward incremental circuit analysis (see Problem P6.4) shows that the short-circuit current gain for this composite connection is approximately the product of the

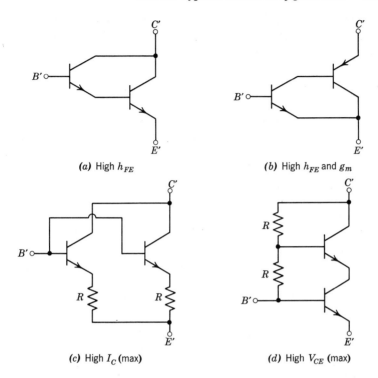

**Fig. 6.2.** Composite transistors ($E'$, $B'$, and $C'$ correspond approximately to the emitter, base and collector of a single transistor).

betas of the two separate transistors. In some instances this composite transistor can be used simply to replace a single transistor, with a corresponding increase in gain.

The circuit in Fig. 6.2b achieves approximately the same high current gain as the Darlington connection, but, in addition, the effective $g_m$ of the composite transistor is greatly increased (see Problem P6.4). This connection can also be used to replace a single transistor when the added gain is desirable.

Figures 6.2c and d show how two transistors can be direct coupled so as to increase the current or voltage capability over that of a single transistor. The resistors are used to insure that the current or voltage is shared equally between the two units (see Problem P6.5).

## 6.1.2 Interstage Coupling

Usually it is necessary to cascade one or more common-emitter stages to achieve sufficient over-all gain. The simplest cascading arrangement is shown in Fig. 6.3a. Although this circuit forces one to use essentially zero quiescent collector-to-base voltage on the first stage, the circuit can operate quite effectively, particularly with silicon transistors: for low values of $I_C$, and $V_{CE} \geq 0.3$ volt, the transistor is essentially in a linear region of amplification. Since $V_{BE}$ is usually at least 0.5 volt, it is clear that satisfactory operation is possible. The disadvantage of this zero-bias operation is that the collector space-charge layer is so narrow that $C_\mu$, $g_\mu$, $\omega_T^{-1}$, and $h_{FE}^{-1}$* are all higher than they would be at, say, several volts of bias. Unfortunately, we cannot achieve these higher bias voltages except at an increase in circuit complexity and cost. The important question is to determine which is "better"; more stages of sub-optimum gain, or fewer stages with more interstage circuitry.

If several volts of collector voltage are required, we might consider a coupling scheme such as shown in Fig. 6.3b. In this circuit, if the current through $R_3$ and $R_4$ is made significantly larger than the base current of the second stage, it is possible for $R_3$ to be small compared with the input impedance of the second stage and still have a significant dc voltage drop across it. Unfortunately, a small value of $R_3$ would require either a small value of $R_4$ or a large value of the negative supply voltage, neither of which is normally desirable. In addition, if the current in $R_3$ is comparable to the collector current of the first transistor, then $R_1$ must be reduced and a significant amount of power is wasted in the bias circuitry. An application where this circuit is useful is one in which $R_3$ can be bypassed for all signal frequencies, so that a large $R_3$ will not imply a large signal attenuation. In this case, $R_4$ can sometimes be returned to ground so that only one supply voltage is required. Note

---

* $h_{FE} \equiv (I_C/I_B)$ $V_{CE}$ constant

### Sec. 6.1 Typical Circuit Configurations

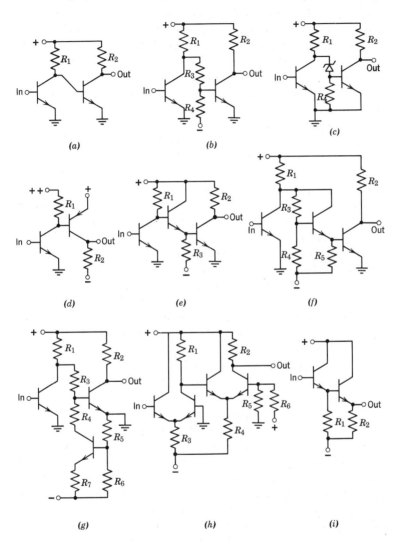

**Fig. 6.3.** Interstage coupling circuits.

that this circuit, with $R_3$ bypassed, offers an improvement over usual ac-coupled amplifiers because the relatively large emitter bypass capacitor is not needed and, as we will see, the operating points can still be stabilized by over-all negative feedback.

Figure 6.3c shows a modification which overcomes most of the disadvantages of the circuit of Fig. 6.3b, though at an increase in cost. Here $R_3$ has been replaced by an avalanche (or Zener) diode which is operated in the avalanche mode with a voltage selected to be approximately equal to the desired quiescent collector-to-base voltage of the first transistor. The necessary current through the diode is often small enough to allow $R_4$ to be connected directly to ground without the necessity of providing a negative supply. One advantage of this circuit is that the increase in avalanche voltage with temperature can be made to compensate for the decrease in base-to-emitter voltage with temperature; the temperature dependence of the avalanche voltage is such that compensation is usually achieved for an avalanche voltage between 5 and 7 volts. One important disadvantage of the avalanche diode as a coupling element is its associated low-frequency noise, which can be quite large. Hence the coupling scheme of Fig. 6.3c may not be usable for input stages of high-gain dc amplifiers. Note that in some instances it may be possible to use one or more forward-biased diodes in lieu of the avalanche diode; the choice here is likely to be governed by cost and temperature considerations.

If both *pnp* and *npn* transistors are available, the circuit of Fig. 6.3d is attractive. Unfortunately, for good design freedom in the choice of resistors and operating points, it is desirable to have three separate power supplies or the equivalent. If we are willing to settle for about 0.6 volt across $R_1$ and $R_2$, then, of course, only a single supply is needed and alternating *pnp* and *npn* stages can be cascaded indefinitely. However, a word of caution is required. Note that there is a direct path from the plus supply voltage through the two transistors to ground; there is virtually nothing to limit the current that can flow in this path. Hence only a few milliamperes of input to the first stage could cause hundreds of milliamperes of collector current with possible resulting damage to the transistors. In fact, a disastrous chain reaction is possible in an amplifier with many stages. The first transistor can heat excessively and form a collector-to-emitter short circuit. This then causes the next stage to heat excessively, and . . . . Hence, it is common to connect a small resistor in series with an appropriate base or emitter lead in order to limit this overload current. Note that in Fig. 6.3 the burnout problem arises only in circuits (d) and (e). In Fig. 6.1c, $R_1$ was included only to prevent burnout. In capacitively coupled amplifiers this burn-out problem does not often arise.

We should certainly not overlook the possibility of using transistors in the coupling network. Figure 6.3e shows one attractive possibility. This configuration provides only about 0.6 volt collector-to-base bias voltage (i.e., $V_{CB} = V_{BE}$) but this relatively small bias can produce significant reduction in $C_\mu$ and an increase in $\omega_T$ compared to the zero-bias operation in Fig. 6.3a, particularly for transistors with a wide near-intrinsic region between base and collector (e.g., transistors made by using the epitaxial process). Moreover, the added transistor causes a significant increase in overall current *and* voltage gain, besides providing improved quiescent conditions for the common-emitter stages. Since this circuit suffers from the overload burn-out problem, a small resistor should be introduced in series with, say, the collector of the common-collector stage.

Figures 6.3f and g show two examples of circuits in which a transistor is used solely as a coupling element and accomplishes nothing that could not be done, in principle, at least, with batteries and resistors. In Fig. 6.3f the coupling transistor simulates the effect of the avalanche diode in Fig. 6.3c. If the collector current of the coupling transistor is significantly smaller than that of the first transistor, then the operating point of the first transistor is determined by $R_1$, $R_3$, and $R_4$ almost exactly as in Fig. 6.3b. However, the coupling transistor, by virtue of its current gain, creates the effect of a small incremental resistance in parallel with $R_3$ (see Problem P6.6). In Fig. 6.3g the coupling transistor operates very nearly as a constant current source (see, for comparison, $T_3$ in Fig. 3.3a) and thereby simulates the effect of a large $R_4$ and a large negative supply voltage (see Problem P6.7).

By using two-transistor pairs as the building block, it is possible to construct useful cascades, such as the one shown in Fig. 6.3h. The values of $R_5$ and $R_6$ can be adjusted to give good design freedom for the choice of operating points. Moreover, this type of cascade can give good frequency response because of the nature of common-collector and common-base stages (see Problem P6.8).

As a last example, note that the cascade of identical common-collector stages is very straightforward, as indicated in Fig. 6.3$i$. This cascade is used where large current gain, but no voltage gain, is required. The circuit can easily be extended to three or four cascaded stages, but eventually the apparent current gain becomes illusory.

### 6.1.3 Examples of Direct-Coupled ac Amplifiers

The circuit configurations discussed in the last section are suitable for achieving very high gains. There will not be any gain, however, unless suitable input biasing is used to maintain the proper operating points for the transistors. If dc amplification is required, we have no choice but to try to devise temperature-compensation schemes which automatically vary the input bias so that the desired signal will not be masked by the effect of changes in temperature. If, however, we do not require dc amplification, it is possible to use large amounts of dc feedback to reduce the dc gain and thereby stabilize the operating points of the transistors. Note, for example, that in the feedback amplifier example in Fig. 3.3$a$, a large capacitor in parallel with $v_F$ would eliminate high-frequency feedback but leave the low-frequency feedback intact; thus the feedback, with $f = 0.05$, would be effective in reducing the output level from saturation to only 160 millivolts. We could even preserve some ac feedback, but if the feedback network is frequency selective, so that there is much less ac feedback than dc feedback, then the ac gain can be very large without creating temperature-drift problems caused by large dc gain. In this section we examine two examples of direct-coupled ac amplifiers.

Figure 6.4 shows two possible circuit configurations for achieving good bias stability through dc feedback while still allowing the desired ac gain. Figure 6.4$a$ is a two-stage common-emitter amplifier with feedback to the input emitter. For frequencies in the passband, the amount of feedback is decreased by $C_2$, which becomes an effective short circuit. The amplifier has a mid-band input impedance of slightly less than 50 k (i.e., $R_1 || R_2$) and a voltage gain

## Sec. 6.1 Typical Circuit Configurations

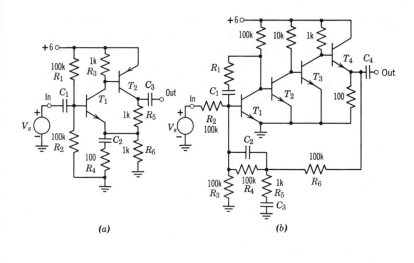

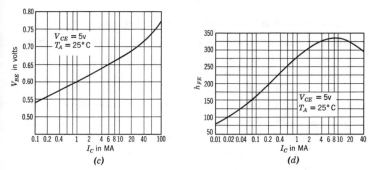

**Fig. 6.4.** Examples of direct-coupled, ac amplifiers. (a) Two stage, positive gain. (b) Four stage, negative gain. (c) $V_{BE}$ versus $I_C$. (d) $h_{FE}$ versus $I_C$.

of 12 (controlled by $R_4$, $R_5$, and $R_6$). Coupling capacitors $C_1$ and $C_3$ prevent the operating point from depending on source and load characteristics.

Figure 6.4b shows a four-stage amplifier with feedback to the input base. This amplifier has essentially zero collector-to-base voltage for the first two transistors. This zero-bias operation allows a relatively simple circuit configuration without any need for a negative supply voltage. The large amount of dc feedback is greatly reduced by $C_3$, which becomes a short circuit in the pass band. For

## 182  Direct-Coupled Amplifiers

ac signals this amplifier has an input impedance of 100 k (i.e., $R_2$), a voltage gain of 100 (controlled by $R_2$, $R_4$, $R_5$, and $R_6$), and an output impedance of less than 1 ohm. High-frequency compensation consists of $R_1$, $C_1$, and $C_2$, which are adjusted to prevent oscillation and eliminate overshoot in the step response caused by the ac feedback. For this example, it has been assumed that there is a dc path through the source, $V_s$, and that a few microamperes of current through $V_s$ is tolerable. We could use an input coupling capacitor, but it would introduce some change in the transistor quiescent operating points unless $R_3$ was reduced to 50 k.

Note that in both of these examples direct coupling, together with dc feedback, has not only reduced the number of coupling and bypass capacitors and bias resistors, but has also reduced low-frequency phase shift in the amplifier. Hence, there is no problem in avoiding low-frequency oscillations caused by the large amount of ac feedback.

### 6.1.4 Calculation of Quiescent Operating Conditions

Let us consider how we might calculate the operating point of a multistage direct-coupled amplifier such as the one in Fig. 6.4a. These same methods will also be applicable to balanced or symmetric amplifiers with high dc gains, as we shall discuss in Sec. 6.3. The key simplifying feature is the fact that if the operating points are reasonably insensitive to temperature, they must also be reasonably insensitive to the exact value of base current and emitter-to-base voltage. Thus a crude *guess* as to $I_B$ and $V_{BE}$ should lead to a fairly accurate determination of $I_C$ and $V_{CE}$, and we can then calculate a more accurate approximation for $I_B$ and $V_{BE}$. The process can be iterated until a sufficiently accurate answer is obtained, but if this calculation does not converge *very* rapidly, the circuit is probably undesirably sensitive to the exact values of $I_B$ and $V_{BE}$.

Let us apply the iterative approach to quiescent operating point calculations for the circuit of Fig. 6.4a. Let us assume that $T = 25°C$ and that $T_1$ is a diffused silicon planar transistor described by the characteristics in Fig. 6.4c and d. For simplicity of this example, assume $T_2$ is identical to $T_1$, except for reversal of signs of voltage and current. From c and d, we note that at 25°C, $T_1$ and $T_2$ have a very high value of $h_{FE}$ and that $|V_{BE}|$ is between about 0.5 and 0.7 volt over the current range of $|I_C| = 10\mu a$ to 10 ma. As a first approximation to calculating the operating point, we might try $|V_{BE}| = 0.6$ volt and $I_B = 0$ for both transistors. Figure 6.5a shows the operating points corresponding to this assumption. The circled numbers indicate a possible sequence in which these values might be calculated.

Using the collector operating points indicated in Fig. 6.5a and the characteristic curves of Fig. 6.4c and d, we can now make a better approximation for $I_B$ and $V_{BE}$. This second approximation for base current and voltage leads to a revised estimate of collector current and voltage as indicated in Fig. 6.5b. Note that our first approximation was really quite good, and the second approximation is certainly close enough to allow us to determine the incremental circuit parameters. Unit-to-unit and temperature-induced variations will probably produce errors greater than the error of the approximate calculation.

A second example of this iterative method is indicated in Fig. 6.6; the transistors are presumed to be of the same type as in the last example. Note that, by starting with the dc feedback network and then working backward through the amplifier, all operating

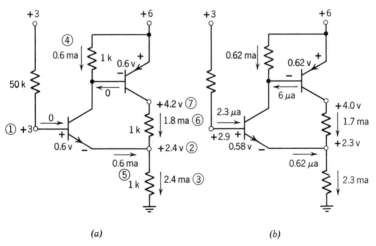

**Fig. 6.5.** Operating point calculation for circuit in Fig. 6.4a. (a) First approximation ($I_B = 0$, $V_{BE} = 0.6$ volt). (b) Second approximation ($I_B$ and $V_{BE}$ were calculated on the basis of $I_C$ and $V_{CE}$ in (a) and the data in Fig. 6.4c and d).

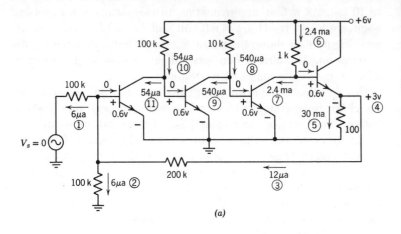

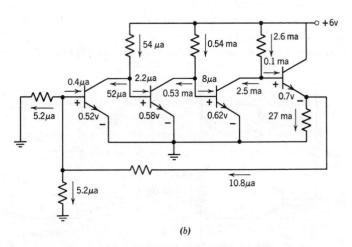

**Fig. 6.6.** Operating point calculation for circuit of Fig. 6.4b. (a) First approximation ($I_B = 0$, $V_{BE} = 0.6$ volt). (b) Second approximation ($I_B$ and $V_{BE}$ were calculated on the basis of $I_C$ and $V_{CE}$ in (a) and the data in Fig. 6.4c and d).

points can be readily calculated. For this example the second approximation is accurate enough for virtually any purpose. Note that self-heating causes a 35°C rise in junction temperature for the fourth stage transistor, which affects slightly the operating points. Note also that, because of the large dc feedback, the voltage at the emitter of $T_4$, $V_{E4}$, is almost exactly five times the base voltage for $T_1$, $V_{B1}$. If $V_{B1}$ were to change about 2.4 mv/°C then $V_{E4}$ would change about 12 mv/°C. Even for a ±30°C change in temperature, the collector current of the fourth transistor would only change ±20%, which is probably good enough for most applications. If better operating-point stability is required, resistor $R_3$ in Fig. 6.4b can be increased in value and returned to a negative supply voltage, or $R_5$ and $R_6$ can have an appropriate value of positive temperature coefficient. Note that, as the circuit stands, the quiescent collector currents of $T_1$ and $T_2$ are practically independent of temperature.

## 6.2 EFFECTS OF PARAMETER CHANGES

### 6.2.0 *Introduction*

One of the most important problems in dc amplifier analysis is to consider properly the effect of changes or uncertainties in circuit parameters. There are three reasons for this importance:

1. The signal circuit and bias circuit are not distinct. Hence, changes in such things as temperature or power supply voltage may produce an effect indistinguishable from a change in signal level.

2. Many dc amplifiers are used for critical applications such as meter amplifiers, oscilloscopes, and analog computers. These applications require rather precise gain and impedance levels in spite of changes in power supply voltage, temperature, and device characteristics.

3. An important class of dc amplifiers uses symmetric circuits to compensate partially for unavoidable parameter changes. The performance of this type of circuit depends critically on small changes from perfect symmetry.

This section treats the problem of predicting the effect of changes in circuit parameters. In particular, the principle interest is in developing an *approximate* method for dealing with *small changes*. Section 6.3 uses this method to consider the problems described above.

### 6.2.1 *Calculation of the Effect of a Change in Transconductance*

Consider first the problem of calculating the effect of a change in the value of transconductance of one element in a linear circuit.

## 186  Direct-Coupled Amplifiers

The situation in question can be depicted as shown in Fig. 6.7a. In this figure all fixed sources are presumed to be external to the box labelled LSC (Linear Stable Circuit) and, in the interest of simplicity, this figure shows only two fixed sources, $V_a$ and $I_b$. Sources $V_a$ and $I_b$ are labeled $F_a$ and $F_b$ to emphasize their *fixed* (or independent) nature. They cause currents to flow in various branches of the circuit and, in particular, cause dependent generator $D_x$ to generate a current $G_x V_x$, where $V_x$ is the voltage at the indicated terminal pair. The problem is to ascertain what changes in currents or voltages would be produced by a change in the coefficient of the dependent generator $D_x$. We are interested in computing the effect of this change on any or all branch voltages and/or currents. We could, of course, compute currents before and after the change, and then compute the difference. But if the change is small, this leads to subtraction of almost equal quantities and, hence, simplifying approximations are difficult to apply. In practice, the change might be very small and still be of great interest; hence, we prefer to develop a method to *compute the change alone*.

First, let the values of terminal voltages and currents after the change be as indicated in Fig. 6.7b. The subscript $o$ is used to designate the "original" value before the change and the prefix $\Delta$ is used to indicate the amount of change. Since we are dealing with a linear, stable circuit we can represent currents and voltages in Fig. 6.7b as the superposition of the original values in Fig. 6.7a plus the changes which are indicated in Fig. 6.7c. In Fig. 6.7c the generator labeled "?" generates a current $\Delta I_x$ which, on the basis of Figs. 6.7a and b, is given by:

$$\Delta I_x = (I_{xo} + \Delta I_x) - I_{xo}$$
$$= (G_{xo} + \Delta G_x)(V_{xo} + \Delta V_x) - G_{xo} V_{xo}$$
$$= \Delta G_x V_{xo} + (G_{xo} + \Delta G_x)\Delta V_x$$
$$= \underbrace{\frac{\Delta G_x}{G_{xo}} I_{xo}}_{①} + \underbrace{(G_{xo} + \Delta G_x)\Delta V_x}_{②} \quad (6.1)$$

We note that term ① in Eq. 6.1 is a quantity which can be computed on the basis of Fig. 6.7a and the known value of $\Delta G_x$; hence,

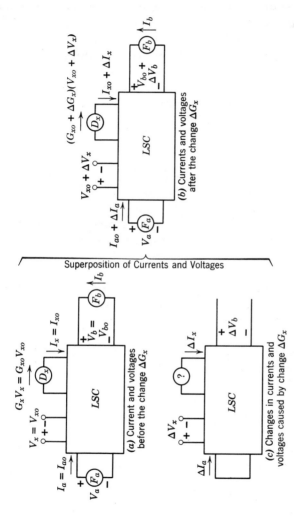

**Fig. 6.7.** Calculation of currents and voltages after a change in transconductance. $LSC$ is a linear stable circuit with no internal sources, $D_x$ is a dependent source, and $F_a$ and $F_b$ are fixed sources. The nature of generator "?" is discussed in the text.

it can be treated as a *known* source as far as currents and voltages in Fig. 6.7c are concerned. Term ②, on the other hand, depends on $\Delta V_x$ which is a quantity we are trying to compute; hence ② is a *dependent* source. Generator "?" can thus be depicted in two parts as shown in Fig. 6.8a.

Figure 6.8a can be interpreted as follows. When the dependent generator coefficient changes from $G_{xo}$ to $G_{xo} + \Delta G_x$, the voltages and currents in the circuit will change. The amount of each *change* is *exactly* equal to the corresponding response produced by the known fixed current source, of value

$$I = \Delta G_x V_{xo} = \frac{\Delta G_x}{G_{xo}} I_{xo}$$

placed in parallel with the changed dependent source $(G_{xo} + \Delta G_x)\Delta V_x$. The original fixed sources (in this case $F_a$ and $F_b$) are set equal to zero in this calculation.

The key idea was recognition of the fact that $V_{xo}$ could be calculated for the original circuit and hence was *known* as far as Fig. 6.8a was concerned. Since the change $\Delta G_x$ was also known, the generator of value $(\Delta G_x)V_{xo}$ became a *fixed* source! Thus the *exact* alterations of currents and voltages produced by the change from $G_{xo}$ to $G_{xo} + \Delta G_x$ could be calculated *directly* from Fig. 6.8a.

Note that the dependent current source could depend on any voltage or current in the circuit. The only things we need to know are the current in the generator before the change, and the fractional change in the coefficient (e.g., $\Delta G_x/G_{xo}$). The special case where $V_x$ equals the voltage across $D_x$ is, of course, the case of a self-conductance and thus we need not discuss this important case explicitly.

For convenience in future discussion we use the term "perturbation generator" to designate the fixed generator which is used to simulate the change (e.g., source $F_x$ in Fig. 6.8). The rest of this section deals with the use of perturbation generators for circuit calculations.

There is, of course, a dual development for analyzing the effects of changes in a dependent voltage source. The perturbation generator then becomes a fixed voltage source in series with the changed dependent voltage source as shown in Fig. 6.8b. By analogy with Fig. 6.8a, the coefficient of the perturbation generator would be

## Sec. 6.2 Effects of Parameter Changes 189

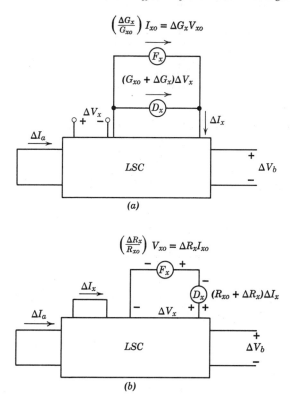

**Fig. 6.8.** Use of perturbation generator, $F_x$, to depict change. (a) Effect of change in dependent current source ($G_{xo}$ changes to $G_{xo} + \Delta G_x$). (b) Effect of change in dependent voltage source ($R_{xo}$ changes to $R_{xo} + \Delta R_x$).

the fractional change in the dependent source coefficient (e.g., $\Delta R_x/R_{xo}$) multiplied by the voltage across the dependent source before the change took place.

The use of a perturbation generator for computing the effect of changes is illustrated in Fig. 6.9. (This circuit is like a transistor circuit with a 5-k series emitter resistance.) The assumed problem is to calculate the effect of a 10% increase in the value of the transconductance. Figure 6.9a shows the initial current distribution prior to the change. Note that initially the dependent generator current is 8.0 milliamperes. To compute the effect of the 10% change, we introduce a 0.8-milliampere (10% × 8.0) *fixed* source

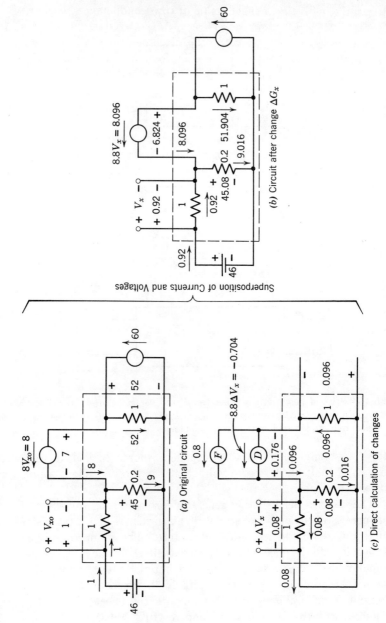

**Fig. 6.9.** Example of calculation of the effect of changes. (Circuit values are volts, milliamperes, and millimhos.)

in parallel with the changed value of transconductance, as shown in Fig. 6.9c, and compute the effect of this 0.8-milliampere source on the rest of the circuit with other fixed sources dead. This is a straightforward, though not trivial, circuit analysis problem, which we need not discuss in detail here. Figure 6.9b shows the total currents after the change.

**6.2.2** *Approximate Analysis of Small Changes*

An important special case of the perturbation generator method is the restriction to *small* changes. The key idea is the recognition that if $\Delta G_x \ll G_{xo}$, then (with reference to Fig. 6.8a) we can make the approximation:

$$(G_{xo} + \Delta G_x)\Delta V_x \cong G_{xo}\Delta V_x \tag{6.2}$$

In short, for small changes, the term $\Delta G_x \Delta V_x$ is a second-order quantity in comparison with the term $G_{xo}\Delta V_x$. Thus, for small changes, we can compute the approximate effect of the change by introducing the perturbation generator into the original circuit *without changing the coefficient of the dependent generator*.

For most transistor circuits, it appears that parameter changes of as much as 10% can be handled adequately by neglecting the second-order terms. For example, in the previously considered example in Fig. 6.9, the effect of the 10% change can be computed with less than 9% error (i.e., 0.9% error in *total* quantities) if we assume the dependent generator in Fig. 6.9c is $8.0\Delta V_x$. Fortunately we do not usually have to compute the effect of the *change* with much greater accuracy than about 10%, so the approximation is adequate. If the precise effect of the change is important, then often the change itself is smaller and the approximation is still valid.

The principal advantage of the approximate method is that it allows the effects of several simultaneous changes to be combined readily. An example which illustrates this fact is shown in Fig. 6.10. Here the circuit is the same as in the previous example, except that the problem of interest is different. Consider the circuit in the dashed box in Fig. 6.10a to be the model for a linear amplifier, and assume that we wish to calculate the change in gain and impedance characteristics of the amplifier which would accompany a simultaneous increase of two circuit elements by 10%. In particular, assume we wish to calculate the change in short-circuit

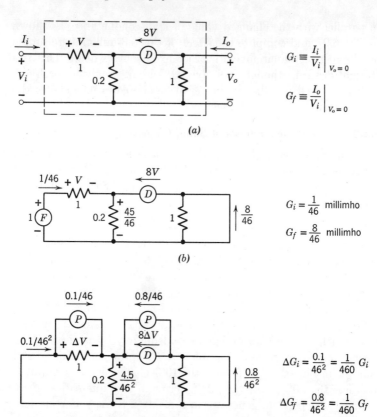

**Fig. 6.10.** Approximate calculation of change in $G_i$ and $G_f$ for simultaneous change of two parameters ($F$ = fixed source, $D$ = dependent generator, $P$ = perturbation generator). Units are volts, milliamperes, millimhos. (a) Amplifier circuit. (b) Amplifier with unit input voltage, short-circuit output. (c) Approximate effect of 10% increase in two elements in circuit.

input conductance $G_i$ and forward transconductance $G_f$ caused by a change in the two upper elements in the circuit (i.e., changes in what is equivalent to $g_\pi$ and $g_m$).

The calculation proceeds as follows. We first connect a unit input source and a short-circuit load, and compute the initial conditions as shown in Fig. 6.10b. From the currents indicated in Fig. 6.10b we conclude that $G_i = 1/46$ millimho and $G_f = 8/46$

millimho. Next, we remove the fixed unit source and introduce perturbation generators for each element which is undergoing a change. According to the approximate method, we leave the circuit unchanged and thus the effects of the two changes can be computed by superposition. We thus calculate that the branch currents change by the amounts indicated in Fig. 6.10c and, hence, $\Delta G_i/G_i = \Delta G_f/G_f = 1/460$. Surprisingly enough, the two 10% changes acting together almost exactly cancel, so there is only about a 0.2% change in $G_i$ and $G_f$.

This numerical example illustrates several important ideas:

1. Although the effect of two simultaneous parameter changes cannot be analyzed *exactly* as the sum of the effects of the two changes acting independently (i.e., the effects of parameter changes do not satisfy superposition), nevertheless for small changes the effects are approximately additive.

2. Small changes in input conductance, current gain, etc., can be calculated approximately by introducing perturbation sources into the original circuit with the parameters unchanged.

3. Several simultaneous changes can be partially (or totally) compensating. In fact, an important objective of dc amplifier design is to cancel any unavoidable change by deliberately introduced compensating changes.

### 6.2.3 *Effect of Small Changes in a Nonlinear Circuit*

The use of perturbation generators to depict change is valid only for linear circuits because the proof depended on superposition However, if the circuit is incrementally linear, the perturbation-generator method can be used for approximate analysis of small changes. The key to the analysis is to approximate the non-linear characteristics in terms of the first-order terms of a Taylor series expansion of the $V$-$I$ characteristics about the desired operating point.

As an example, suppose a diode voltage $V_D$ is presumed to be a function of current $I_D$ and temperature $T$. Assume that the characteristics are as shown in Fig. 6.11a and that we know the operation will be in the vicinity of point "$Q$". If we neglect all but the first-order terms, we are, in effect, assuming that the characteristics, near $Q$, can be approximated by a series of parallel, uniformly

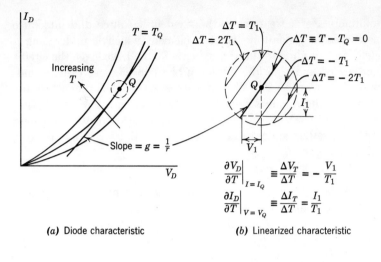

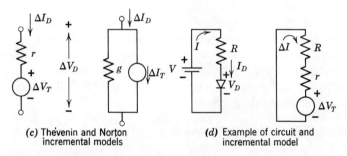

(c) Thévenin and Norton incremental models

(d) Example of circuit and incremental model

**Fig. 6.11.** Calculation of effect of a small change in a non-linear circuit.

spaced lines, as shown in Fig. 6.11b. For circuit analysis we can use either the Thévenin or Norton circuit model as shown in Fig. 6.11c; the quantities $I_D$ and $V_D$, $r$, and $g$ are defined in Fig. 6.11a, and $\Delta V_T$ is defined in Fig. 6.11b. The use of these models is indicated in Fig. 6.11d for a simple series circuit. In this circuit a change in temperature causes a change in $I$ given by

$$\Delta I = -\frac{\Delta V_T}{R + r}$$

This example considered only the effect of a change in temperature, but of course other changes could be handled equally well. If several changes occur simultaneously, we can replace $\Delta V_T$ or

$\Delta I_T$ by the sum of several terms, each determined by a different change-producing mechanism. The important conclusion is that the effect on voltage and current of small changes in additional parameters (like temperature) of a nonlinear, two-terminal $V$-$I$ characteristic can be analyzed by means of a linear Thévenin or Norton model. Note specifically that *only* a voltage or current source is introduced by the change; any change in the incremental resistance is not a first-order effect as far as changes in dc voltage and current are concerned.

Note, however, that there are some circuits which utilize directly the change in impedance as a first-order effect. For example, the bias current can be varied in order to vary the incremental impedance of a diode, and hence control the gain of an amplifier. In this case, however, the *impedance* is the quantity of direct interest, so the Taylor-series expansion should be for the *impedance*, not for a current or voltage.

### 6.2.4 Representation of Changes in Transistor Characteristics

The methods discussed in the last section are readily applicable to depicting change in transistor characteristics. The only difference is that it requires two independent voltage and/or current variables to characterize a transistor. Thus, if we wish to depict the effect of a change in temperature we need to add two temperature-dependent generators. *In short, the effect of small changes in the characteristics of a transistor can be completely represented by two perturbation generators located at the available terminals.*

For example, if we use admittance notation, the effect of temperature changes upon the incremental voltages and currents in a transistor can be represented as follows:

$$\Delta I_B = g_i \Delta V_{BE} + g_r \Delta V_{CE} + \Delta I_{By}$$
$$\Delta I_C = g_f \Delta V_{BE} + g_o \Delta V_{CE} + \Delta I_{Cy} \tag{6.3}$$

where $\Delta$ refers to changes from nominal quiescent values and $\Delta I_{By}$ and $\Delta I_{Cy}$ are temperature-controlled generators for the "y" representation. Clearly, $\Delta I_{By}$ and $\Delta I_{Cy}$ are given by:

$$\Delta I_{By} = \frac{\partial I_B}{\partial T}\bigg|_{V_{BE}, V_{CE}} \Delta T$$
$$\Delta I_{Cy} = \frac{\partial I_C}{\partial T}\bigg|_{V_{BE}, V_{CE}} \Delta T \tag{6.4}$$

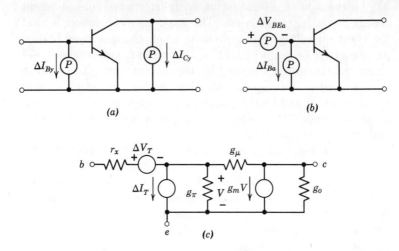

**Fig. 6.12.** Possible choices for perturbation generators to characterize the effect of small changes in transistor dc characteristics. (a) Two current sources (assuming $V_{BE}$ and $V_{CE}$ are fixed). (b) Two input sources (assuming $I_C$ and $V_{CE}$ are fixed). (c) Incremental model with $r_x$ external to $\Delta V_T$ and $\Delta I_T$.

where the $V_{BE}$, $V_{CE}$ notation means these variables are held fixed. Inclusion of the temperature effect as perturbation generators is illustrated schematically in Fig. 6.12a. Of course, in actually using this representation, the transistor should be replaced by its usual *incremental model*.

A somewhat more convenient choice of perturbation generators is shown in Fig. 6.12b. This model treats $V_{CE}$ and $I_C$ as independent variables which are held constant while the change takes place. The generators $\Delta V_{BEa}$ and $\Delta I_{Ba}$ then signify the amount of increase in $V_{BE}$ and $I_B$ that would be required to hold $I_C$ and $V_{CE}$ fixed.

For temperature-induced changes, the values of the perturbation sources in Fig. 6.12b are:

$$\Delta V_{BEa} = \left.\frac{\partial V_{BE}}{\partial T}\right|_{V_{CE},I_C} \Delta T \qquad (6.5)$$

$$\Delta I_{Ba} = \left.\frac{\partial I_B}{\partial T}\right|_{V_{CE},I_C} \Delta T \qquad (6.6)$$

## Sec. 6.2 Effects of Parameter Changes

The subscript $a$ is used to indicate that the *output* variables are held constant, as in the $ABCD$, or "general circuit constant" description of networks. The advantage of the model in Fig. 6.12$b$ is that since all changes are "referred to the input" they can be compared directly with input signal levels. This is desirable because the effect of the perturbation sources is indistinguishable from changes in signal levels and we must insure that the effect of the signal is not masked by the effect of the change.

To express the perturbation generators in terms of the quiescent conditions on the transistor, $I_{CQ}$, $V_{CEQ}$, $T_Q$, we use a relation of the familiar form*

$$I_C = AT^3 \exp\left(\frac{V_{BE} - V_{go}}{kT/q}\right) \quad \text{(signs are for an } npn \text{ unit)} \tag{6.7}$$

where $A$ is a constant depending on the type and size of transistor, $V_{go}$ depends on the material properties, and $k$ and $q$ are the usual physical constants. Equation 6.7 is accurate for the intrinsic transistor, but does not include effects of base resistance; thus we must represent the base resistance as external to the perturbation generators, as shown in Fig. 6.12$c$. For this reason we have redefined the generators in Fig. 6.12$c$ as $\Delta V_T$ and $\Delta I_T$. With $r_x$ considered external, Eq. 6.7 is valid at low to moderate values of $I_C$ and is capable of very good accuracy if $I_{CO}$ is small. It is significant that $A$ is the only parameter which depends significantly on transistor dimensions; hence, this parameter alone is likely to vary from one transistor to another of the same type.

To find $\Delta V_T$, we solve Eq. 6.7 for $V_{BE}$ and take the partial derivative with respect to temperature, as in PEM, Sec. 3.3.4. The result is

$$\Delta V_T = \left.\frac{\partial V_{BE}}{\partial T}\right|_Q \Delta T \cong \frac{\Delta T}{T_Q}[V_{BEQ} - V_{go}] \tag{6.8}$$

where, as in PEM, we have dropped a small term $3kT_Q/q$ compared to $(V_{BEQ} - V_{go})$. For silicon $npn$ transistors $V_{go} = 1.205$ volts. Typically, $\Delta V_T$ is between $-2.2$ and $-2.7$ millivolts per °C.

* See PEM, Secs. 9.1.3 and 3.3.4.

The expression for $\Delta I_T$ in Fig. 6.12c cannot be calculated with the same precision as $\Delta V_T$. We can, however, express it in a form that is convenient for analysis:

$$\Delta I_T = \frac{\partial I_B}{\partial T}\bigg|_Q \Delta T = \frac{\partial}{\partial T}\left(\frac{I_C}{h_{FE}}\right)\bigg|_Q \Delta T = \frac{I_{CQ}}{(h_{FEQ})^2}\frac{\partial h_{FE}}{\partial T}\Delta T$$

$$= I_{BQ}\left[\frac{1}{h_{FEQ}}\frac{\partial h_{FE}}{\partial T}\right]_Q \Delta T \qquad (6.9)$$

Equation 6.9 expresses $\Delta I_T$ in terms of the fractional change in $h_{FE}$. For silicon *npn* transistors, at moderate current levels and near room temperature, $h_{FE}$ increases about 1% per °C increase in temperature. Thus, if $I_{BQ} = 100$ nanoamperes, then $\Delta I_T$ will be about 1 nanoampere per °C.

### 6.2.5 Three Types of Change

It is important to appreciate that several types of change are significant in dc amplifier operation and each type poses a distinct circuit design problem. In this section three types of changes are discussed, with particular mention of their implications in dc amplifier analysis and design.

First, consider the effect of changes or uncertainties due to manufacturing variations. Suppose that we design a circuit on the basis of *average* characteristics for each of the circuit elements. When we then build the circuit we might imagine that each *physical* component is actually an *average* one connected to an appropriate number of perturbation generators. Thus the dc voltages and currents in the circuit will differ from their design values by some amount which is presumably small and could be computed from knowledge of the perturbation generators. In a dc amplifier we usually must compensate for the differences between design and reality. In particular, with zero input we may require precisely zero output. Note that if we consider a two-port amplifier as a whole, the differences between design values and actual values can be described by two perturbation generators in the same way that changes in a transistor can be described by two perturbation generators. Hence *two* adjustments are necessary to nullify the effects of the perturbation generators. One possible choice of adjustments, usually called "zero" or "balance" adjustments, is to

## Sec. 6.2 Effects of Parameter Changes    199

introduce an adjustable voltage source in series with the input and an adjustable current source in shunt with the input. These sources are then adjusted so as to buck out or nullify the perturbation generators associated with the manufacturing variations in the amplifier.

As a practical matter, the zero adjustments usually must be at or near the amplifier input. To see why, note that although the perturbation generators must be small compared to the total voltage or current, they need not be small compared with the incremental or signal values. Thus, for example, the deviation between design values and actual values may be enough to cause the output stage of the amplifier to saturate even with zero input. For illustration, refer to the example in Chapter 3, Fig. 3.3, and note that for $f = 0$ violent asymmetries would have to be introduced into the *output* stage in order to provide zero output for zero voltage input. For contrast, calculations show that (in Fig. 3.3) only a $+15\%$ change in one $R_1$ and a $-15\%$ change in the other $R_1$ could produce the same effect. We should not try to compensate for manufacturing variations by means of zero adjustments in the output circuit alone and, in general, the design of proper zero adjustment circuitry is a major problem in itself.

A second important type of change is that which is associated with a measurable environmental change. The typical example is a change or "drift" associated with changes in temperature or power supply voltage. Since the environmental change is "measurable," we can derive a voltage or current which is proportional to the change-producing mechanism; then we can feed this voltage and/or current into the circuit in such a way as to compensate for any first-order changes in circuit behavior. Thus, we might use a transistor as a temperature sensor; the variation of the emitter-to-base voltage with temperature could be used to compensate for some unavoidable changes of circuit performance with temperature.

In principle, *any small change caused by a measurable effect can be compensated for*, but often at considerable expense. Moreover, the relation between the change and the cause of the change can sometimes be established only on the basis of measurement, and every amplifier must then be compensated individually. Some adjustments, such as compensation for power supply changes, can be made easily because the power supply voltage can (usually) be changed easily. Others, such as temperature compensation, may require time-consuming measurements in a controlled environment; the process of compensation could require several hours for each amplifier. It is *preferable*, though not always possible, to *design* an amplifier to be temperature-compensated as closely as possible without the need to make extensive experimental adjustments. Note, particularly, that there can be an interaction (i.e., a mutual dependence) between zero adjustments, temperature compensation adjustments, etc.

A third important class of change is "unpredictable" change. Thermal and shot noise are important examples in this category. Also, aging effects in the transistors and resistors may fall in this category. Admittedly, for some applications we are willing to make periodic adjustments to compensate for gradual changes in circuit characteristics, but in other applications this is not possible. The best design procedure for minimizing long term aging or "drift" is to use circuit components with a high degree of stability. Thus wire-wound resistors are preferable to carbon resistors, and silicon planar transistors are preferable to germanium alloy units. It is important to distinguish between this long-term, essentially unpredictable change and short-term changes which are more often predictable effects caused by changes in temperature or power supply voltage.

Another important type of unpredictable change is what is usually called "flicker" or "$1/f$" noise.* This manifests itself as a low-frequency fluctuation, not too different from shot noise or thermal noise. The frustrating aspect of flicker noise, however, is that its magnitude is essentially independent of bandwidth for a dc amplifier. In ac amplifiers with lower half-power frequencies greater than about 1000 cps, the important noise is primarily thermal or shot noise, and the total noise power output from the amplifier can be reduced by decreasing the bandwidth. If, however, the power density spectrum has *approximately* a $1/f$ frequency dependence and if the bandwidth extends to dc, then reducing the upper half-power frequency will have little effect on noise power output. For example, if a dc amplifier with a 10 cps bandwidth has a peak-to-

---

* See Chapter 4 of *Characteristics and Limitations of Transistors*, by R. D. Thornton, D. DeWitt, E. R. Chenette, and P. E. Gray, hereafter referred to as CLT.

peak flicker noise of about 1 microvolt, then the use of shunt capacitors to reduce the bandwidth to 0.1 cps will not significantly alter this 1 microvolt value. The proper choice of operating point, the use of selected low-noise transistors, and operation in a stable temperature environment with balanced, temperature-compensated circuits can minimize flicker noise. As a practical matter, the best that can be achieved with currently available transistors is an apparent base-to-emitter low-frequency noise on the order of a few tenths of a microvolt peak-to-peak.

## 6.3 SYMMETRIC CIRCUITS

### 6.3.1 Some General Principles

The use of symmetry in high-gain dc amplifiers is probably the single most important technique for minimizing undesirable effects caused by predictable changes. This section will illustrate some of the important ideas relating to symmetrical circuits.

Two identical common-emitter stages can be interconnected in a symmetrical fashion as indicated in Fig. 6.13. Note that if the connection between emitters is removed, then each stage has an operating point which can be well stabilized by proper choice of resistor values (see ECP, Chapter 5). If the two sides of the circuit are identical, then when the circuits are connected as shown, there will be no change in operating point. There will, however, be a significant increase in gain between $V_s$ and $V_{\text{out}}$ because the degeneration (or negative feedback) caused by the series emitter resistance is nullified. To see this, consider the incremental model shown in Fig. 6.13b. In the interest of simplicity, this model assumes $r_x = g_o = g_\mu = 0$. The balanced nature of this circuit makes it impossible for $V_s$ to cause current through $R_e$; we can visualize the circuit as a balanced bridge driven by $V_s$, and with $R_e$ at the balance point. Note that if $V_s$ can not cause current to flow through $R_e$, then the voltage across $R_e$ will not change and we can, in effect, replace $R_e$ with a short circuit. Alternatively, we might think of the symmetry of the circuit as accomplishing the same effect that is accomplished by a bypass capacitor across $R_e$ in an ac-coupled amplifier.

It is not difficult to see that an important advantage of an ideal symmetrical circuit, such as Fig. 6.13, is that if $V_{\text{in}} = 0$, then

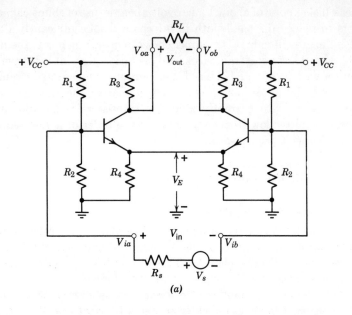

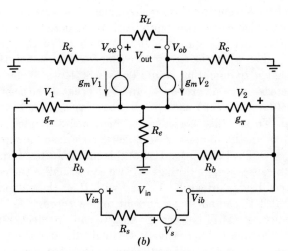

**Fig. 6.13.** Symmetry applied to a common-emitter amplifier

$$(R_1 || R_2 = R_b,\ R_3 = R_c,\ R_4 = 2R_e)$$

(a) Circuit. (b) Incremental model.

### Sec. 6.3 Symmetric Circuits   203

$I_{in} = 0$ and $V_{out} = 0$, and hence, the output is insensitive to changes in supply voltage or temperature. The extent to which this advantage can be achieved is usually determined by unavoidable asymmetries between the two halves, particularly asymmetries in the dc characteristics of the active devices, and in the temperature dependence of all of the circuit elements. A major difficulty with the circuit of Fig. 6.13 is that the input circuit, the output circuit, and the power-supply circuit are isolated from each other and do not have a common ground; this problem will be considered later in more detail.

Note that a symmetric dc amplifier can be constructed by combining any two identical, direct-coupled ac amplifiers. For example, the ac direct-coupled amplifier of Fig. 6.4a can be converted into the symmetric dc amplifier as shown in Fig. 6.14, with an interconnection taking the place of bypass capacitors. The quiescent operating points are unchanged (assuming perfect symmetry) and the dc voltage gain is the same as the mid-band ac gain of the original amplifier. The input and output impedances are doubled because the two amplifiers are, effectively, in series. This strong similarity between symmetric circuits and direct-coupled, ac amplifiers allows us to greatly simplify the analysis. The quiescent operating point can be found by open-circuiting all symmetric connections that carry no quiescent current (e.g., the 200-ohm resistor in Fig. 6.14) and analyzing half of the remaining circuit in the manner described in Sec. 6.1.4. The dc gain is then determined by analyzing a "mid-band" incremental model. For signal analysis, any voltage which is not changed by the signal (e.g., $V_E$ in Fig. 6.13) is a virtual ground. Thus, for example, the mid-point of the 200-ohm resistor in Fig. 6.14a is a virtual ground, and the gain, input impedance, etc., can be found by analyzing the half-circuit shown in Fig. 6.14b.

Some types of symmetric circuits cannot be analyzed as two interconnected nonsymmetric circuits without recourse to an ideal 1:1 transformer. For example, the circuit of Fig. 6.15a has cross-coupling resistors that constitute positive feedback. If the circuit is in a symmetric condition with equal collector currents, then the currents in the two 1M resistors must be equal, so we can calculate the quiescent operating conditions on the basis of Fig. 6.15b. For incremental gain calculations the model of Fig. 6.15c must be used.

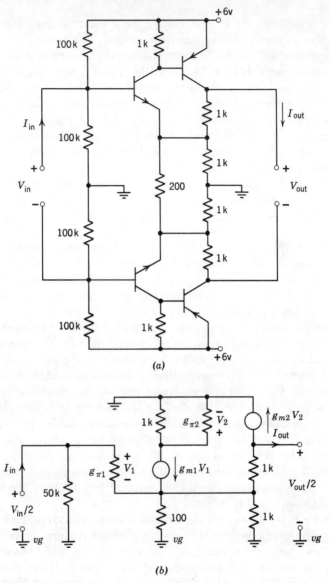

**Fig. 6.14.** Symmetric circuit derived from Fig. 6.4a. Incremental transistor parameters are determined for the quiescent operating conditions given in Fig. 6.5b. A virtual ground caused by symmetry is designated $vg$. (a) Complete circuit. (b) Incremental model for signal analysis.

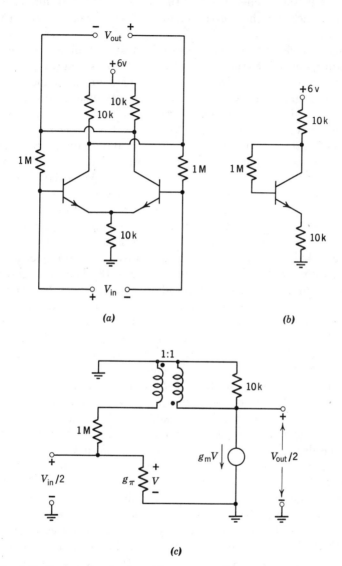

**Fig. 6.15.** Symmetric circuit with cross-coupling. (a) Complete circuit. (b) Bias half-circuit. (c) Signal half-circuit.

Note that this model requires the use of a 1:1 transformer in order to achieve the correct phase for the current in the 1M resistor.

Figure 6.15a can also be used to illustrate a problem that can arise in symmetric circuits. Note that this circuit strongly resembles a flip-flop, or bistable circuit. If the source and load impedances are both high it may well turn out that the symmetric state, with both transistors in their normal ranges with identical operating points, is not stable and Fig. 6.15b will lead to erroneous results.

In general, one must remember that merely finding a self-consistent operating point does not, in any way, insure that this point represents a stable condition. This problem is particularly severe in symmetric circuits because it is possible for the circuit to be stable in the difference mode but unstable in the common mode, or vice versa. Thus in feedback amplifiers one must always consider whether what looks like difference-mode negative feedback may not also be common-mode positive feedback. For example, if we cascade two stages of the amplifier of Fig. 6.13, then we might be tempted to apply negative transadmittance feedback by placing two resistors from output collectors to symmetrically opposite input bases. Such a scheme does, in fact, lead to negative difference-mode feedback because the difference-mode half circuit will contain a 1:1 phase reversing transformer. However, if the emitter resistances (i.e. $R_e$ in Fig. 6.13) are not high enough, the resulting common-mode *positive* feedback can cause the circuit to behave as a four-transistor flip-flop with two undesirable stable states.

The possibility of undesirable stable states is not, of course, limited to symmetric circuits, and is a problem which must always be considered in all multistage direct-coupled circuits. Note that even if the desired state *is* stable, we may still also have undesirable stable states. Consequently, a circuit may be able to "lock" into an undesired stable state even though the desired state is also stable (if it could be reached!). Fortunately, these problems are usually not difficult to solve. For example, one might add a diode which would conduct if the circuit tried to jump into an undesired stable state. If the diode is connected at the right place in the circuit, it can eliminate the undesired stable state but not affect the desired stable one. It is thus not unusual for a dc amplifier to contain a diode which *never* conducts!—or to be more precise, *almost* never conducts.

### 6.3.2 Analysis by Means of Sum and Difference Components

The previous discussion has indicated the possibility of analyzing symmetric circuits by considering various half-circuits. This technique is, in fact, quite general and more powerful than the previous examples have indicated. Moreover, the method can be extended to the analysis of almost symmetric circuits.

Consider, first, a two terminal-pair linear circuit which might also contain independent sources, as shown in Fig. 6.16a, and is described by:

$$I_a = G_{aa}V_a + G_{ab}V_b + I_{aq}$$
$$I_b = G_{ba}V_a + G_{bb}V_b + I_{bq}$$
(6.10)

$G_{aa}, G_{ab}, G_{ba}, G_{bb}$ are incremental conductances. $I_{aq}, I_{bq}$ are quiescent currents.

If $G_{aa} \cong G_{bb}$, and $G_{ab} \cong G_{ba}$, then it is sometimes more convenient to solve not for $I_a$ and $I_b$ in terms of $V_a$ and $V_b$ but for $I_a + I_b$ and $I_a - I_b$ in terms of $V_a + V_b$ and $V_a - V_b$. The necessary algebra is quite straightforward. For convenience, we start by defining the "common-mode" and "difference-mode" quantities:

$$V_c \equiv \tfrac{1}{2}(V_a + V_b) = \text{common-mode voltage}$$
$$V_d \equiv \tfrac{1}{2}(V_a - V_b) = \text{difference-mode voltage}$$
$$I_c \equiv \tfrac{1}{2}(I_a + I_b) = \text{common-mode current}$$
$$I_d \equiv \tfrac{1}{2}(I_a - I_b) = \text{difference-mode current}$$
(6.11)

Equations 6.11 may also be written as:

$$V_a \equiv V_c + V_d \qquad V_b \equiv V_c - V_d$$
$$I_a \equiv I_c + I_d \qquad I_b \equiv I_c - I_d$$
(6.12)

On rewriting Eqs. 6.10 in terms of the common-mode and difference-mode quantities, we have:

$$I_c = G_{cc}V_c + G_{cd}V_d + I_{cq}$$
$$I_d = G_{dc}V_c + G_{dd}V_d + I_{dq}$$
(6.13)

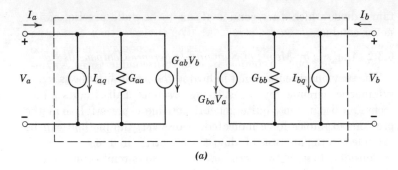

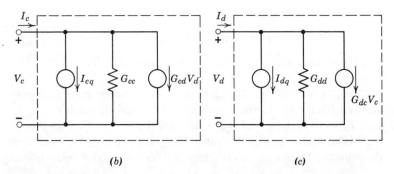

**Fig. 6.16.** Analysis of two-port by means of common-mode and difference-mode half-circuits. See Eqs. 6.10 to 6.14 for definition of terms. (a) Circuit representation; for left-right symmetry $G_{aa} = G_{bb}$ and $G_{ab} = G_{ba}$, but $I_{aq}$ need not equal $I_{bq}$. (b) Common-mode half-circuit; for symmetry $G_{cd} = 0$. (c) Difference-mode half-circuit; for symmetry $G_{dc} = 0$.

where

$$G_{cc} = \tfrac{1}{2}(G_{aa} + G_{ab} + G_{ba} + G_{bb}) \quad G_{cd} = \tfrac{1}{2}(G_{aa} - G_{ab} + G_{ba} - G_{bb})$$

$$G_{dc} = \tfrac{1}{2}(G_{aa} + G_{ab} - G_{ba} - G_{bb}) \quad G_{dd} = \tfrac{1}{2}(G_{aa} - G_{ab} - G_{ba} + G_{bb})$$

$$I_{cq} = \tfrac{1}{2}(I_{aq} + I_{bq}) \qquad\qquad I_{dq} = \tfrac{1}{2}(I_{aq} - I_{bq})$$

(6.14)

The inverse relations for $G_{aa}$, $G_{ab}$, $G_{ba}$, and $G_{bb}$ are given by Eqs. 6.14 with the interchange of subscripts $a \leftrightarrow c$, and $b \leftrightarrow d$. The two half-circuits corresponding to Eqs. 6.13 are shown in Fig. 6.16b and c.

If $G_{aa} = G_{bb}$ and $G_{ab} = G_{ba}$, then $G_{cd} = G_{dc} = 0$, and Eqs. 6.13 simplify to two *uncoupled* equations. The circuit is then said to be symmetric. Note that the fixed sources are not symmetric; the symmetry refers only to conductances or transconductances. A symmetric circuit can be analyzed as two *independent* half-circuits, as indicated in Fig. 6.16b and c with $G_{cd} = G_{dc.} = 0$; one half-circuit is used for determining common-mode currents and voltages (e.g., $I_c$, $V_c$) and the other half-circuit is used for determining difference-mode currents and voltages (e.g., $I_d$, $V_d$).

Figure 6.17 illustrates the use of half-circuits for analyzing symmetric circuits. Analysis of the complete circuit requires the solution of a pair of simultaneous equations, while the common-mode and difference-mode half-circuits require only a single equation each. We can see by inspection of Fig. 6.17b that $I_c = 3$ and from Fig. 6.17c that $I_d = 1$. Thus (using Eqs. 6.12) $I_a = (I_c + I_d) = 4$ and $I_b = (I_c - I_d) = 2$. In this example there is no significant simplification produced by the use of common-mode

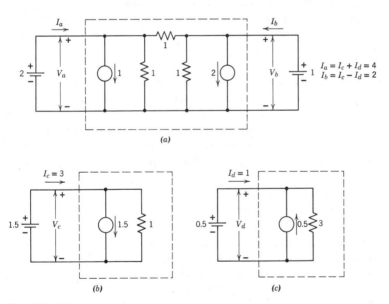

**Fig. 6.17.** Example of analysis by sum and difference currents (units of mhos, volts, amperes). (a) Complete circuit. (b) Common-mode half-circuit. (c) Difference-mode half-circuit.

and difference-mode circuits, but for more involved circuits the degree of simplification can be enormous.

The common-mode circuit can be found directly without any algebraic manipulations. We simply assume $V_a = V_b$ and remove any branches which carry no current. It may also be necessary to swap two leads which carry the same current, such as in Fig. 6.15$b$. To find the difference-mode circuit, we assume $V_a = -V_b$ and make a ground connection to any point which is unaffected by the presence of $V_a$ and $V_b$. In some cases a 1:1 transformer is also needed, as in the example of Fig. 6.15$c$. Note, again, that the independent sources in the network need not be constrained by symmetry requirements. It is only the incremental resistances and transconductances (or equivalent) which must have a symmetrical relation.

As a second example of the use of sum and difference components, consider the symmetric, common-emitter amplifier shown in Fig. 6.18. This circuit has two inputs and two outputs and is often called a differential amplifier because the gain for difference-mode signals is much greater than the gain for common-mode signals. The circuit is commonly used when a desired difference-mode signal is superimposed on a large common-mode signal. Figure 6.19 shows one possible application where the differential amplifier is used with a bridge circuit and an oscilloscope to measure a slight unbalance caused by a small change in resistance, $\Delta R$. It is assumed that the oscilloscope and power supply, $V$, *must* have a common terminal. Thus it is important that the oscilloscope respond *only* to the difference-mode component of the input, $(V_{ia} - V_{ib})/2$, and not to the common-mode component, $(V_{ia} + V_{ib})/2$.

Returning, then, to the circuit of Fig. 6.18 we must first find the quiescent operating points. Next, a suitable incremental model must be found; in this case it is assumed that $g_\pi$ and $g_m$ are the only important parameters, but $g_\mu$ and $g_x$ may be important in other situations. Using the incremental model shown in Fig. 6.18$b$, we can draw the common-mode and difference-mode half-circuits as shown in Fig. 6.18$c$ and $d$. The analysis is then very similar to Eqs. 6.10 to 6.14 except that we are dealing here with only the *incremental* model (Fig. 6.18$b$), which has no fixed sources that are independent of signal levels $V_{ia}$ and $V_{ib}$. Thus terms analogous to

Sec. 6.3 Symmetric Circuits 211

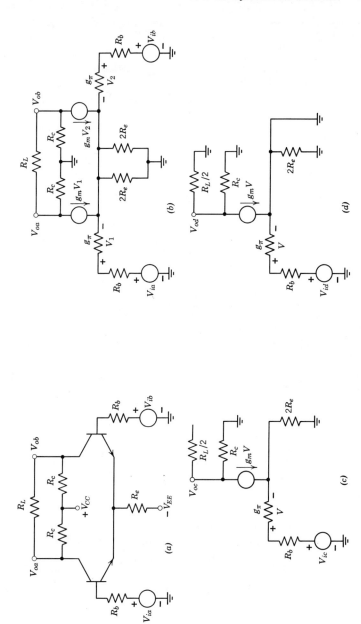

**Fig. 6.18.** Analysis of common-emitter differential amplifier by means of common-mode and difference-mode half-circuits. (a) Complete circuit. (b) Incremental model. (c) Common-mode half-circuit. (d) Difference-mode half-circuit.

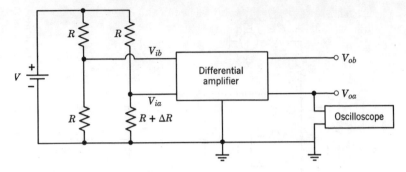

**Fig. 6.19.** Differential amplifier used to detect bridge unbalance.

$I_{aq}$ and $I_{bq}$ in Eq. 6.10 will be absent here. Accordingly we find the common-mode and difference-mode output voltages in terms of the voltage gains and input voltages:

$$V_{oc} = A_{cc}V_{ic} + A_{cd}V_{id}$$
$$V_{od} = A_{dc}V_{ic} + V_{dd}V_{id} \tag{6.15}$$

We know from the circuit symmetry that $A_{cd} = A_{dc} = 0$ (i.e., a common-mode input can produce only a common-mode output and a difference-mode input can produce only a difference-mode output) and a relatively straightforward calculation shows that, from Fig. 6.18c

$$A_{cc} = \left.\frac{V_{oc}}{V_{ic}}\right|_{V_{id}=0} = \frac{-\beta_0 R_c}{R_b + r_\pi + (\beta_0 + 1)2R_e} \tag{6.16a}$$

and from Fig. 6.18d

$$A_{dd} = \left.\frac{V_{od}}{V_{id}}\right|_{V_{ic}=0} = -\frac{\beta_0(R_c\|R_L/2)}{R_b + r_\pi} \tag{6.16b}$$

If we want to express the separate outputs in terms of the separate inputs, we use relations analogous to Eq. 6.12 and find:

$$V_{oa} = A_{aa}V_{ia} + A_{ab}V_{ib} \tag{6.17a}$$
$$V_{ob} = A_{ba}V_{ia} + A_{bb}V_{ib} \tag{6.17b}$$

where:

$$A_{aa} = \tfrac{1}{2}(A_{cc} + A_{cd} + A_{dc} + A_{dd}) \quad \text{etc.} \tag{6.18}$$

For this example, it is considerably easier to calculate $A_{aa}$, $A_{ab}$, etc. by sum and difference components than by direct analysis. Moreover, we are often interested specifically in $A_{cc}$ and $A_{dd}$ even though $A_{dd}$ may be much greater than $A_{cc}$ (or vice versa). Unless we use extreme care it may turn out that apparently reasonable approximations will amount to neglecting $A_{cc}$ in Eqs. 6.18. The reason for being particularly interested in $A_{cc}$ and $A_{dd}$ separately is that the common-mode component of $V_i$ can be much larger than the difference-mode component or vice versa. Thus the common-mode component of $V_o$ can be quite substantial, even if $A_{cc}$ is orders of magnitude smaller than $A_{dd}$. For example, if the bridge in Fig. 6.19 is balanced, a 1-volt change in $V$ will produce a 0.5-volt common-mode input to the amplifier. The nonzero common-mode gain means that the meter reading will change just as though the bridge had become unbalanced; the usefulness of the amplifier as a null detector is thus impaired.

### 6.3.3 *Effect of Small Asymmetries*

A practical circuit is never perfectly symmetric and it is important to know how small asymmetries will affect the performance. If there is asymmetry, then the exploitation of symmetry to develop *independent* half-circuits is not possible. We can, of course, use coupled half-circuits, but we can no longer exploit symmetry to calculate $A_{cc}$, $A_{dd}$, etc. The calculations then become so complex that there is little, if any, advantage in using half-circuits. However, if the asymmetry is small, say, less than about 10%, then perturbation generators can be employed to greatly simplify the analysis. The key idea is as follows. Imagine that initially the circuit is perfectly symmetric with parameter values equal to the average of the actual values for each half-circuit. First, compute the output, assuming that the input acts on this perfectly symmetric circuit. Next, assume that a change takes place which increases a parameter on one side of the circuit and decreases the corresponding (or "homologous") parameter on the other side of the circuit. If the changes are small, say, less than 10%, then we can approximate the effect of the change by introducing appropriate perturbation generators into an otherwise unperturbed circuit (i.e., we neglect the second-order effects in accordance with the discussion in Sec. 6.2).

## 214   Direct-Coupled Amplifiers

Consider an example to illustrate the use of perturbation generators in the analysis of almost symmetric circuits. Suppose that, in Fig. 6.18$b$, $g_\pi$ for the left transistor is increased by $\Delta g_\pi$ and that $g_\pi$ for the right transistor is decreased by $\Delta g_\pi$. Let us compute the effect of this change on the four common-mode and difference-mode gain expressions $A_{cc}$, $A_{cd}$, $A_{dc}$, and $A_{dd}$. In particular, we should expect that the terms $A_{cd}$ and $A_{dc}$, which are zero for a perfectly symmetric circuit, should now have some nonzero value. Thus, because of the asymmetry, a common-mode input signal might produce a difference-mode output signal, and vice versa. In fact, as we will see, the *only* first-order effect of the perturbation generators will be to alter the cross-coupling terms, $A_{cd}$ and $A_{dc}$, and $A_{cc}$ and $A_{dd}$ are not affected by small asymmetries. The steps in the analysis are indicated in Fig. 6.20 and will now be explained.

The calculation indicated in Fig. 6.20 is done in two parts. First, the original symmetric circuit is excited by a difference-mode signal as indicated in Fig. 6.20$a$(1), and the current through the eventually asymmetric element, $g_\pi$, is determined. Suitable perturbation generators are then connected, as shown in Fig. 6.20$a$(2); the minus sign associated with the right generator occurs because $g_\pi$ for that transistor is presumed to decrease by $\Delta g_\pi$. The two fixed perturbation generators in Fig. 6.20$a$(2) clearly excite *only* the common mode in what is still a perfectly symmetric circuit. Hence, the only effect of the perturbation generators is to produce a common-mode output, and the value of this output can be deduced from the common-mode half-circuit shown in Fig. 6.20$a$(3). The detailed calculations show that:

$$I_d = \frac{V_d}{R_b + r_\pi} \qquad \text{[from Fig. 6.20}a\text{(1)]}$$

$$\frac{V_{oc}}{(\Delta g_\pi/g_\pi)I_d} = \frac{\beta_0 R_c (R_b + 2R_e)}{R_b + r_\pi + (\beta_0 + 1)2R_e} ; \qquad \text{[from Fig. 6.20}a\text{(3)]}$$

Thus

$$A_{cd} = \left.\frac{V_{oc}}{V_{id}}\right|_{V_{ic}=0} = \frac{V_{oc}}{V_d}$$

$$= \frac{\Delta g_\pi}{g_\pi} \frac{\beta_0 R_c (R_b + 2R_e)}{(R_b + r_\pi)(R_b + r_\pi + (\beta_0 + 1)2R_e)}$$

In a similar fashion we can find $A_{dc}$ from Fig. 6.20b:

$$I_c = \frac{V_c}{R_b + r_\pi + (\beta_0 + 1)2R_e} \quad \text{[from Fig. 6.20}b(1)\text{]}$$

$$\frac{V_{od}}{(\Delta g_\pi/g_\pi)I_c} = \frac{\beta_0 R_b(R_c \| R_L/2)}{R_b + r_\pi} \quad \text{[from Fig. 6.20}b(3)\text{]}$$

Thus

$$A_{dc} = \left.\frac{V_{od}}{V_{ic}}\right|_{V_{id}=0} = \frac{V_{od}}{V_c}$$

$$= \frac{\Delta g_\pi}{g_\pi} \frac{\beta_0(R_c \| R_L/2)R_b}{(R_b + r_\pi)(R_b + r_\pi + (\beta_0 + 1)2R_e)}$$

Of course, asymmetries in other elements than $g_\pi$ will also affect $A_{cd}$ and $A_{dc}$. In fact, asymmetries can be self-cancelling, and we could even deliberately introduce a *controllable* asymmetry to cancel the effect of an uncontrolled one. Thus we might experimentally adjust $R_c$ for one transistor so that a common-mode input produces *no* difference-mode output; the procedure, of course, is equivalent to balancing a bridge and can be done with great precision. Unfortunately, this asymmetry in $R_c$ might adversely affect other parameters such as sensitivity to power supply or temperature fluctuations.

### 6.3.4 *Effect of Power-Supply Variations*

One important reason for using symmetry is to reduce the effect of variations in power-supply voltage. For example, typically, a *good* power supply will have a voltage which changes on the order of 0.02% per °C change in temperature; so, for a 10-volt supply and a 10°C change this becomes a 20-millivolt change. Without the use of symmetry this 20-millivolt change might cause, typically, a change in output equivalent to a 1-millivolt input. Suppose, however, that the output is taken between the two collectors in Fig. 6.18. If the circuit were perfectly symmetric, the output would not change *at all* if $V_{CC}$ or $V_{EE}$ changed. The incremental gains and impedances would change somewhat, but, if necessary, over-all feedback could be employed to reduce these changes. The short-coming of the symmetric circuit is apparent when we admit that asymmetry

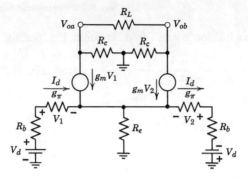

(1) Symmetric circuit with $V_d$ input

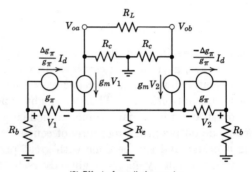

(2) Effect of small change $\Delta g_\pi$

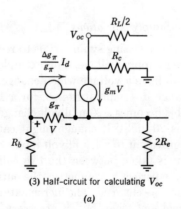

(3) Half-circuit for calculating $V_{oc}$

(a)

**Fig. 6.20.** Calculation of effect on incremental gain of increasing $g_\pi$ of left transistor to $g_\pi + \Delta g_\pi$ and decreasing $g_\pi$ of right transistor to $g_\pi - \Delta g_\pi$. (a) Difference mode excitation.

## Sec. 6.3 Symmetric Circuits 217

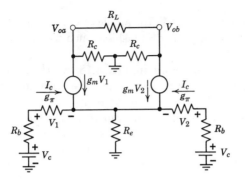

(1) Symmetric circuit with $V_c$ input

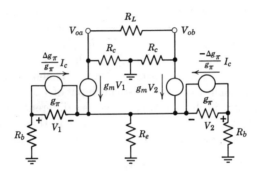

(2) Effect of small change $\Delta g_\pi$

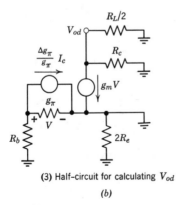

(3) Half-circuit for calculating $V_{od}$

(b)

**Fig. 6.20.** (*Continued*) (*b*) Common-mode excitation.

exists and, even worse, when the application requires that the output be taken in an asymmetric fashion, such as from one collector to ground.

Consider, first, the circuit in Fig. 6.18, which is presumed to exhibit perfect symmetry, and assume we wish to consider the effect of $V_{CC}$ and $V_{EE}$ changing by amounts $\Delta V_{CC}$ and $\Delta V_{EE}$. These changes are represented in the linear incremental model of Fig. 6.21.

From Fig. 6.21 we see that $\Delta V_{CC}$ will produce *only* a common-mode output and, in fact, no amount of asymmetry would cause a difference-mode output due to this source. In actual fact, of course, the transistor is not perfectly unilateral and we should include the conductances $g_\mu$ and $g_o$ in the model. With $g_\mu$ and $g_o$ properly accounted for, the source $\Delta V_{CC}$ could cause a significant difference-mode output. *Thus it may be necessary to include these parameters in the model even if they have a negligible effect on the gain.*

Source $\Delta V_{EE}$ in Fig. 6.21 can also produce a common-mode output and, in fact, a straightforward analysis shows that $\Delta V_{EE}$ is exactly equivalent to a common-mode input of $V_{ic} = \Delta V_{EE}$. Hence, any asymmetry in the circuit could cause a difference-mode output. The fact that $\Delta V_{EE}$ may be much larger than the difference-mode input signal requires that $A_{cd}$ be very much less than $A_{dd}$.

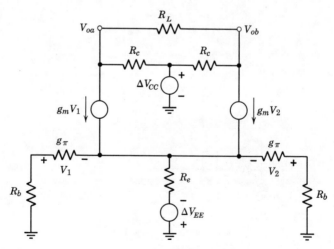

**Fig. 6.21.** Linear incremental model for calculating effect of power supply variations.

## Sec. 6.3 Symmetric Circuits 219

If $\Delta V_{CC}$ and $\Delta V_{EE}$ change together in exactly the right ratio, their effects on common-mode output can be made to cancel. Thus one useful circuit design technique is to derive $V_{CC}$ and $V_{EE}$ from "identical" supplies and choose $R_e$ and $R_c$ such that equal amounts of $\Delta V_{CC}$ and $\Delta V_{EE}$ produce no net common-mode output.

### 6.3.5 Effect of Temperature Variations

The effects of variations in temperature on symmetric or almost-symmetric circuits can be analyzed by using the temperature-dependent generators discussed in Sec. 6.2.4. Take as an example the circuit shown in Fig. 6.22a. The incremental model, with temperature-dependent sources included, is shown in Fig. 6.22b. Because we are assuming that the circuit is driven from a low-impedance source (in this case a voltage source), the effect of the $\Delta I_T$ generators is going to be much smaller than that of the $\Delta V_T$ generators (recall that typical values are $\Delta V_T = -2.5$ mv/°C, and $\Delta I_T = 1$ nanoampere/°C for $I_{BQ} = 100$ nanoamperes). Thus to make the circuit as insensitive to temperature as possible, we want

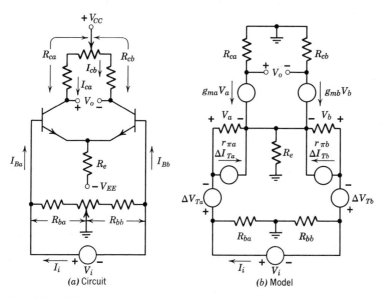

**Fig. 6.22.** Difference amplifier with self-compensation of temperature sensitivity.

to make $\Delta V_{Ta} = \Delta V_{Tb}$. From Eq. 6.8, *this can be achieved by operating the transistors with equal base-to-emitter voltages*, i.e. $V_{BEa} = V_{BEb}$. If the transistors are not perfectly matched, this may require that the collector currents be slightly unbalanced. To achieve the desired zero offset (zero output for zero input) under these conditions, the collector resistors will have to be slightly unbalanced.

The two conditions of equal $V_{BE}$ and zero offset can be realized in practice as follows. First, with shorted input, set the output $V_o$ to zero by adjusting the collector resistors $R_{ca}$ and $R_{cb}$. Second, with open input, reset $V_o$ to zero by adjusting the base resistors $R_{ba}$ and $R_{bb}$. The procedure is then repeated until balance is achieved.

To calculate the thermal drift, we assume the two emitter currents are

$$I_{Ea} = I_{sa} e^{qV_{BEa}/kT}$$
$$I_{Eb} = I_{sb} e^{qV_{BEb}/kT} \quad (6.19)$$

If the two potentiometers have been adjusted as above, then, because the offset is zero,

$$\frac{R_{ca}}{R_{cb}} = \frac{\alpha_{Fb} I_{Eb}}{\alpha_{Fa} I_{Ea}} \quad (6.20)$$

and, because the base voltages have been set equal,

$$\frac{R_{ba}}{R_{bb}} = \frac{(1 - \alpha_{Fb}) I_{Eb}}{(1 - \alpha_{Fa}) I_{Ea}} \quad (6.21)$$

Also,

$$\frac{g_{ma}}{g_{mb}} = \frac{I_{Ca}}{I_{Cb}} \cong \frac{I_{Ea}}{I_{Cb}} \quad (6.22)$$

Under these conditions, and assuming that $\Delta I_T$ does not contribute to the temperature dependence, direct circuit analysis will show that $V_o$ will not be affected by temperature-induced changes in $V_{BE}$; assuming that both transistors are at the same temperature. Also, incremental power supply variations will be cancelled out in $V_o$. (See Problem P6.9.)

In practice, this circuit would not possess perfect thermal drift cancellation, because there are other detailed effects which imply that $\Delta V_{Ta}$ will not be exactly equal to $\Delta V_{Tb}$ when $V_{BEa} = V_{BEb}$. Also, we neglected the effect of the $\Delta I_T$ generators because of the low-impedance drive, but this effect is not zero. It is possible, by adding two more adjustments to the amplifier, to reduce the thermal drift still further by using a slight unbalance in $V_{BE}$ to compensate for all other thermal drifts in the amplifier.*

## PROBLEMS

**P6.1** Figure 6.23 shows the circuit for a meter amplifier. The input current, $I_i$, is amplified by the two-stage amplifier and produces a much larger meter current, $I_m$. The following questions are concerned with the design and analysis of this circuit.

(a) Assume that the transistors are of the type described in Fig. 6.4c and d, and that $R_1 = 50$ k, $R_2 = 2$ k, $R_4 = 20$ k, $R_m = 1$ k. Calculate the values of $R_5$ and $R_6$ which would give the operating points $I_{C1}$ (first stage) = 50 $\mu a$, $I_{C2}$ (second stage) = 500 $\mu a$ at $T = 25°$ C.

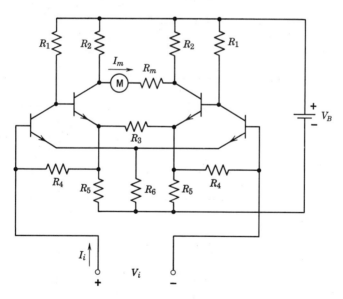

Fig. 6.23. Meter amplifier.

* "Limitations of Transistor DC Amplifiers," A. H. Hoffait and R. D. Thornton, *Proc. of IEEE*, February, 1964.

(b) For the conditions of (a), compute the value of $R_3$ required to make $I_m/I_i = 100$. For this choice, what is the incremental input resistance, $\Delta V_i/\Delta I_i$. (Assume that the transistor can be modeled by an incremental circuit consisting of $g_\pi$ and $g_m$ and that $kT/q = 25$ millivolts.)

(c) For the conditions of (b), compute the expected temperature-induced drift, $\partial I_m/\partial T$, for both open-circuit and short-circuit input. Assume that at 25°C the circuit is perfectly symmetric, but that there is asymmetry in the temperature-controlled generators for the input transistors in the amount 10 $\mu$v/°C and 1 na/°C. Neglect any drift due to other circuit components. (*Hint:* See Problem P3.5b.)

**P6.2** The dc amplifier circuit shown in Fig. 6.24 is indicative of circuits commonly used for direct-coupled operational amplifiers (see Problems P3.5

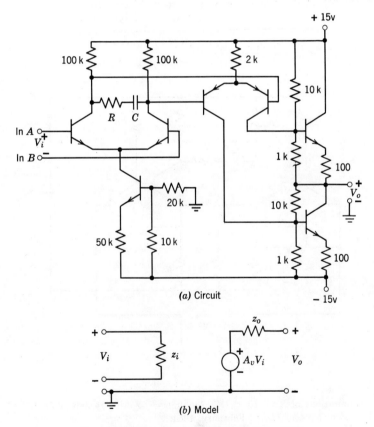

Fig. 6.24. Direct-coupled operational amplifier.

Problems    223

and P4.11). The following questions have to do with analyzing the behavior of this circuit. Assume that all six transistors have characteristics similar to those given in Fig. 6.4c and d, with suitable sign changes for the *PNP* units.

(a) Calculate the quiescent operating points, assuming one input is grounded and $V_i$ is adjusted to have $V_o = 0$.

(b) For difference-mode signals, estimate the input and output impedances and the open-circuit voltage gain; that is, the parameters in Fig. 6.24b.

(c) For common-mode signals, Fig. 6.24b is clearly not valid. What factors would you have to consider in order to estimate the input and output impedances and the open-circuit voltage gain for common-mode signals?

**P6.3** In Fig. 6.20, calculate the values of $A_{cd}$ and $A_{dc}$, assuming the circuit is perfectly symmetric except for the value of $g_m$. Assume that the left transistor has a transconductance $g_m + \Delta g_m$ and the right transistor has a value $g_m - \Delta g_m$.

**P6.4** For the circuits in Fig. 6.2a and b, calculate the short-circuit current gain and transconductance and compare these values for the two circuits. Assume that each transistor can be modeled by using only $g_\pi$ and $g_m$, but do not assume identical parameters since the input transistor will operate at lower collector currents.

**P6.5** What factors govern the choice of $R$ in Figs. 6.2c and d? How do these resistors affect the current gain and transconductance of the composite transistor?

**P6.6** Show that for incremental gain calculations the circuit in Fig. 6.3f can be drawn in the form of Fig. 6.3b. Calculate the values of $R_1$, $R_3$, and $R_4$ for Fig. 6.3b that are needed to make the circuit equivalent to Fig. 6.3f.

**P6.7** The transistor current source shown in Fig. 6.3g can be approximated by an ideal current source in parallel with a resistance. Calculate the values of the current and resistance assuming the transistor can be modeled by $r_\pi$, $g_m$, $r_\mu$, and $r_o$.

**P6.8** Calculate the voltage gain of one stage of the two-stage amplifier shown in Fig. 6.3h. Assume that the series emitter resistance is very large, and that the transistors are identical and can be described by a model containing only $g_m$ and $g_\pi$. What capacitors in the model will have the major effect on high-frequency voltage gain and what is the approximate upper half-power frequency?

**P6.9** For the circuit in Fig. 6.22, prove the assertion made in Sec. 6.3.5 that, under the conditions assumed there, the output voltage $V_o$ will not be affected by thermally-induced changes in $V_{BE}$ (assuming that both transistors are at the same temperature).

# 7

# Tuned Multistage Amplifiers

## 7.0 INTRODUCTION

Single-stage tuned amplifiers were discussed in ECP, Chapter 8. In that discussion, the design problems associated with impedance levels, inconvenient parameter values, and instability were considered. When several tuned-amplifier stages are connected in cascade, all of these problems exist, and one or two new ones arise. In particular, the feedback which couples the output to the input in each stage causes the tuning of each stage to depend on the tuning of all the other stages. If this interaction is strong, it makes the tuning or aligning of the stages difficult, and it makes the alignment rather sensitive to variations in transistor parameters. These problems are considered in the sections that follow, and techniques for overcoming them are presented.

As with the single-stage amplifiers discussed in ECP, the gain characteristic of multistage amplifiers should be as flat as possible in the pass band of the amplifier, and the phase characteristic should be as nearly linear as possible. Also, the variations that do occur in the frequency characteristics should be symmetrical with respect to the center of the passband.

## 7.1 CASCADED NEUTRALIZED STAGES

The principal difficulties that arise in the study of multistage tuned amplifiers are associated with the feedback that exists from output to input in each stage. Therefore, the introduction to the subject is considerably simplified by examining first the case in which the transistors have been neutralized* by one of the methods discussed in ECP. After developing the properties of this ideal amplifier, we will see how the presence of some reverse transmission alters its performance.

A small-signal model for a multistage tuned amplifier with neutralized transistors is shown in Fig. 7.1a. For simplicity, the transistors are represented by boxes in this model; the details of the transistor model are shown in Fig. 7.1b. The transistors are

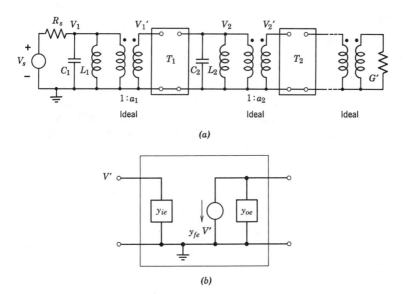

**Fig. 7.1.** Small-signal model for cascaded neutralized stages. (a) Amplifier. (b) Detail for each transistor $T_1$, $T_2$, etc.

*Circuits in which the neutralizing element $y_n$ is adjusted to cancel $y_{re}$ approximately over the pass band of interest are said to be *neutralized*. Circuits in which $y_n$ has a frequency characteristic that *exactly* matches the characteristic of $y_{re}$ at all frequencies are said to be *unilateralized*, because the reverse transmission will then be zero at all frequencies.

characterized by the short-circuit admittance parameters, and the parameters in Fig. 7.1b are composite parameters *including* the small effects of the elements used to neutralize the transistor. For simplicity in this chapter, we shall use the symbol $y$ to represent both the unmodified transistor parameters (i.e., no neutralization) and the composite parameters when neutralization is employed. It will turn out that in fact there is numerically very little difference between these two sets of parameters (except for $y_{re}$, of course). Also, the bias resistors are usually so large that they have a negligible effect on the behavior of the circuit; but if this is not the case, their effects can be lumped with the input admittance $y_{ie}$ in each stage.

The ideal transformers in the circuit of Fig. 7.1a are intended to represent any lossless impedance-transforming network; the actual transformers might be any of the arrangements discussed in Chapter 8 of ECP, such as tapped coils. The voltage relation for the first transformer in the cascade is:

$$V_1' = a_1 V_1 \tag{7.1}$$

and a similar relation holds for each of the other transformers.

To examine the frequency characteristics of the amplifier we consider, first, the gain of the individual stages. Following the procedure used in Chapter 8 of ECP, we can evaluate the voltage gain of the first stage in the vicinity of resonance as:

$$A_{v1}(j\omega') = \frac{V_2'}{V_1'}$$

$$= \frac{-a_2 y_{fe}}{j\omega'[2C_2 + C_{oe} + C_{oe}' + a_2{}^2(C_{ie} + C_{ie}')] + g_{oe} + a_2{}^2 g_{ie}} \tag{7.2}$$

where $\omega'$ is the frequency deviation from $\omega_0$, the center of the pass band:

$$\omega' = \omega - \omega_0 \tag{7.3}$$

and the $y$-parameter capacitances and conductances *for the neutralized transistor* are defined by the equations:

$$y_{oe} = g_{oe} + j\omega_0 C_{oe} + j(\omega - \omega_0)C_{oe}' \tag{7.4}$$

$$y_{ie} = g_{ie} + j\omega_0 C_{ie} + j(\omega - \omega_0)C_{ie}' \tag{7.5}$$

These are the Taylor series expansions of the admittance parameters around $\omega_0$, keeping only one term of the real part and two terms of the imaginary part.

To simplify the notation, we define

$$G_a = g_{oe} + a_2{}^2 g_{ie} \tag{7.6}$$

$$C_a = C_2 + \frac{C_{oe} + C'_{oe}}{2} + a_2{}^2 \left(\frac{C_{ie} + C'_{ie}}{2}\right) \tag{7.7}$$

On this basis, $A_{v1}$ becomes:

$$A_{v1}(j\omega') = \frac{-a_2 y_{fe}}{G_a + 2j\omega' C_a} \tag{7.8}$$

and the bandwidth for a single stage is:

$$\text{Bandwidth} = \frac{G_a}{C_a} \tag{7.9}$$

as derived in ECP, Eq. 8.16.

The voltage gain of each stage in the cascade has the same algebraic form as Eq. 7.8, and the voltage ratio $V_1'/V_s$ at the tuned input also has this form. The over-all voltage gain is the product of these individual gains.

The simplest way to obtain the desired narrow-band over-all gain characteristic is to adjust the individual stages to have the same center frequencies and the same half-power bandwidths. This adjustment is sometimes referred to as *synchronous tuning*. Because the stages are *noninteracting*, the over-all voltage gain for $n$ tuned circuits has the form:

$$A_v(j\omega') = \frac{K}{\left[1 + j\omega'\left(\dfrac{2C_a}{G_a}\right)\right]^n} \tag{7.10}$$

where $K$ is a complex constant for frequencies in or near the narrow pass band of the amplifier. Because Eq. 7.10 is identical in form to Eq. 1.22, the bandwidth for the over-all amplifier will be, by analogy with Eq. 1.25:

$$\text{Over-all bandwidth} = \frac{G_a}{C_a}\sqrt{2^{1/n} - 1} \cong \frac{G_a}{C_a}\sqrt{\frac{\ln 2}{n}} = \frac{0.833\, G_a/C_a}{\sqrt{n}}$$

for large $n$ $\qquad$ (7.11)

(Recall that, in contrast to this result, we noted in Section 5.2 that for *interacting* stages the over-all bandwidth was related to what we called the "per-stage bandwidth" by the factor of $1/n$.)

In general, tuned amplifiers are required not only to amplify signals having frequencies lying in a specified pass band, but also to block the transmission of signals having frequencies lying outside the specified band. To examine the response of the amplifier to signals lying outside the pass band, we note that for frequencies well away from $\omega_0$, $\omega'$ will be large, and $|A_v|$ becomes, from Eq. 7.10:

$$|A_v| = \frac{|K|}{\left(\omega' \dfrac{2C_a}{G_a}\right)^n} \qquad (7.12)$$

Thus the larger the value of $n$, the faster the gain falls off as the signal frequency moves away from the pass band. That is, as the number of tuned circuits is increased, the ability of the amplifier to reject signals lying outside the pass band is increased, and we say that the selectivity of the amplifier is increased.

The theoretical performance of the multistage tuned amplifier using neutralized transistors is quite satisfactory, and the design procedure is straightforward. The relations developed above, together with those developed in Chapter 8 of ECP, can be used to design such amplifiers to meet reasonable specifications. However, there are practical aspects of neutralization that raise some serious problems. Because of the production-line spread of transistor parameters, the neutralizing elements may have to be adjusted individually for each transistor, and this fact hinders production. Also, the neutralization may be impared by changes in temperature or in quiescent operating point. Therefore, it is important to examine what happens to the performance of the amplifier if the neutralization is not perfect, and to consider ways of obtaining satisfactory performance without neutralizing. These matters are considered in the next section.

## 7.2 SINGLE-STAGE BUILDING BLOCK FOR MULTISTAGE AMPLIFIER WITHOUT NEUTRALIZATION

### 7.2.0 *Introduction*

One useful way of studying the effect of imperfect neutralization is to examine the effect on the amplifier pole-zero diagram of

## Sec. 7.2 Single-Stage Building Block for Multistage Amplifier

letting the feedback admittance $y_{re}$ depart from zero simultaneously in each transistor in the amplifier. For perfectly unilateral transistors (i.e., $y_{re} = 0$ at all frequencies) a multistage amplifier with $n$ synchronously tuned LC circuits will have an $n^{th}$ order conjugate pole pair, as shown in Fig. 7.2a. As the magnitude of $y_{re}$ departs from zero, the $n^{th}$-order poles split into $n$ distinct poles, clustered around their original locations. The separation of these poles increases as the magnitude of $y_{re}$ increases, and the amount of the separation can be taken as a measure of the interaction among the tuned circuits in the amplifier. A typical pole-zero diagram for an amplifier having one transistor and two tuned circuits is shown in Fig. 7.2b. As the feedback admittance in this amplifier increases, one pole moves downward and toward the imaginary axis, while the other moves upward and away from the imaginary axis. The skewed manner in which these poles separate leads to skewed frequency characteristics, a condition that has been shown undesirable in Chapter 8 of ECP. Moreover, as $y_{re}$ is increased further, one of the poles moves into the right half of the complex plane, and self-sustaining oscillations result. Thus the effect of the

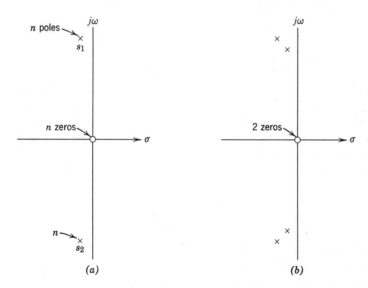

**Fig. 7.2.** Pole-zero diagram for the voltage gain of a multistage amplifier: (a) Unilateral transistors ($y_{re} = 0$), $n$ synchronously-tuned circuits. (b) Transistor with $y_{re} \neq 0$, and two tuned circuits.

feedback is to degrade the performance of the amplifier. In order to insure *stability* and *symmetrical frequency characteristics*, it is necessary to design the amplifier so that these feedback effects are kept small.

In Sec. 7.1 we discussed a synchronously tuned, neutralized amplifier. To obtain the same symmetrical narrow-band characteristics from an amplifier which does *not* have unilateral stages, it is necessary to adjust the various tuned circuits to compensate for the effects of the interaction among the stages. As a practical matter, this alignment of the tuned circuits must be done experimentally, and serious practical problems may arise. When interaction among the stages is present, the adjustment of each tuned circuit depends on the adjustment of the other tuned circuits, so that adjusting one stage may destroy the adjustment of the others. In fact, when the interaction is strong, it will be impossible to achieve the desired synchronous tuning. Thus again we find that we must design tuned multistage amplifiers so that the interaction among the stages is suitably small, this time to insure alignability.

### 7.2.1 *Review of Single Stage Considerations*

Further useful understanding of interaction, stability, and alignment, and of the interrelations among these phenomena, can be gained from the input-admittance locus developed in Sec. 8.3 of ECP. For a brief review of the relations involved, the small-signal model for an amplifier with one transistor and two tuned circuits is shown in Fig. 7.3a. This circuit will ultimately be used to represent each stage in a multistage amplifier. The parameters in this circuit have been transferred across the impedance transformers to the input and output terminals of the transistor.

The feedback current source $y_{re}V_2$ acts as an admittance; this admittance is given by:

$$Y_1 = -\frac{y_{fe}y_{re}}{Y_2} \qquad (7.13)$$

Thus the input circuit for the amplifier can be represented as shown in Fig. 7.3b, with $Y_1$ given by Eq. 7.13. The admittance connected in parallel with the tuned circuit $L_1$, $C_1$ in Fig. 7.3b is:

$$Y_D = G_s + g_{ie} + j\omega_0 C_{ie} + Y_1 \qquad (7.14)$$

## Sec. 7.2 Single-Stage Building Block for Multistage Amplifier

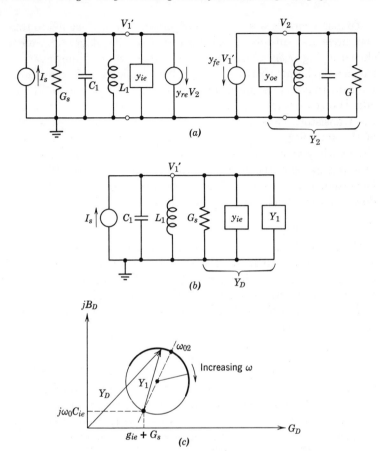

**Fig. 7.3.** Input admittance of a tuned amplifier.

if we assume $y_{ie}$ to be constant near $\omega_0$, (that is, if we assume $C'_{ie} = 0$.) The locus of this admittance as a function of frequency is a circle, as shown in Fig. 7.3c. The radius of the circle is:

$$b = \frac{|y_{fe} y_{re}|}{2(g_{oe} + G)} \tag{7.15}$$

The frequency $\omega_{02}$ indicated in Fig. 7.3c is the resonant frequency of the admittance $Y_2$ at the output of the amplifier, and the portion of the circle shown in a heavy line in Fig. 7.3c corresponds to the band of frequencies between the half-power points for $Y_2$.

To insure circuit stability, the amplifier must be designed so that the circle lies entirely in the right-half of the plane. The locus shown in Fig. 7.3c corresponds to a stable amplifier; however, its frequency characteristics are poor. The proportions in this figure indicate that a substantial part of $Y_D$ is attributable to $Y_1$, the admittance reflected into the input circuit by the feedback through the transistor. Thus $Y_D$ varies substantially over the pass band of the amplifier, and the combination of $Y_D$ and the input tuned circuit will give rise to skewed frequency characteristics. To avoid serious skewing of the characteristics, the radius of the admittance locus should be made small compared with $g_{ie} + G_s$, so that $Y_D$ does not vary much over the pass band.

### 7.2.2 Alignability

Equations 7.13, 7.14, 7.15 and the admittance locus of Fig. 7.3c show that any changes in the parameters in the output circuit affect the tuning of the input circuit by changing $Y_D$. It follows that to keep the interaction between the output and input circuits small, the radius of the circle must be kept small in comparison with $g_{ie} + G_s$. The ratio

$$k = \frac{b}{g_{ie} + G_s} = \frac{|y_{fe} y_{re}|}{2(g_{ie} + G_s)(g_{oe} + G)} \qquad (7.16)$$

is a good measure of the interaction between the output and input, and it proves to be a very useful design parameter. In particular, $k$ gives a good indication of the difficulty that is often encountered in aligning the tuned circuits in multistage amplifiers, and it is therefore called the *alignability factor* for the stage. For small interaction and good alignability, $k$ must be made small. Values of $k$ less than 1/5 are commonly used in well-designed amplifiers. The circle diagram of Fig. 7.3c indicates that when $k$ is made small enough to provide good alignability, the stability of the amplifier and the symmetry of the frequency characteristics are taken care of automatically.

One direct way to obtain a small value for $k$ is to reduce the magnitude of $y_{re}$ by neutralizing. When the neutralization is perfect in the vicinity of the pass band, $k = 0$, and the result is the amplifier discussed in Sec. 7.1. With this adjustment, the interstage networks can be adjusted to match $g_{ie}$ to $G_s$ and $g_{oe}$ to $G$ to achieve maximum gain. However, for reasons presented at the end of Sec. 7.1, neutralizing to obtain small interaction presents some practical problems, and it often does not provide the best means of achieving small interaction.

It is also clear from Eq. 7.16 that $k$ can be made small by making $G$ and $G_s$ large. This adjustment can be obtained easily by proper design of the impedance-transforming networks used to couple $G$ and $G_s$ to the stage, and it turns out that this adjustment is less critical and less sensitive to parameter variations than is the case with neutralization. Thus, in this respect, it may be preferable to obtain small interaction by making $G$ and $G_s$ large rather than by neutralizing. However, making $G$ and $G_s$ large usually precludes the possibility of matching $G$ and $G_s$ to the transistor parameters, so for a given alignability factor, this procedure usually provides less gain than can be obtained by neutralizing and matching impedances. Thus the decision concerning whether to mismatch or neutralize depends primarily on the amount of gain that must be sacrificed by mismatching.

When impedances are to be mismatched to obtain small interaction, it is convenient to define an interstage mismatch factor $m$ equal to the interstage load conductance divided by the interstage driving conductance, calculated on the basis of no interaction, i.e., $y_{re} = 0$. Thus in this case we have, from Fig. 7.3a,

$$m_i = \frac{g_{ie}}{G_s} \quad \text{and} \quad m_o = \frac{G}{g_{oe}} \tag{7.17}$$

The alignability factor can then be written as:

$$k = \frac{|y_{fe}y_{re}|}{2g_{ie}g_{oe}(1 + 1/m_i)(1 + m_o)} \tag{7.18}$$

When a value for the alignability factor has been chosen, Eq. 7.18 can be used to determine the required mismatch factors, and hence the required values for $G$ and $G_s$.

## 7.3 MULTISTAGE AMPLIFIERS WITH INTERACTION

### 7.3.0 *Introduction*

In multistage amplifiers with interaction, the problem of alignability is considerably more complicated than in the single-stage

**234    Tuned Multistage Amplifiers**

prototype discussed in the preceding section, because now we have $n$ intercoupled tuned circuits instead of just two. In making calculations, it is necessary, in principle, to account for interaction between stages by determining the admittance $Y_1$ reflected from the output of each transistor to the input. However, since we must try to keep this effect small, we can achieve considerable simplification in the analysis by first taking steps to guarantee small interaction, and then neglecting the interaction in all subsequent calculations. That is, we choose a small alignability factor for each stage, then determine the required mismatch factors by assuming negligible interaction. This approximation is compensated for when the amplifier is aligned experimentally.

### 7.3.1 Mismatch Factors

Figure 7.4 shows the interstage networks for the first two stages. The representation is valid for the resonant frequency at which the reactive elements cancel out, if the small effect of the interaction is neglected. The mismatch factor at the input of the first transistor is, from Eq. 7.17:

$$m_1 = \frac{a_1^2 g_{ie}}{G_s'} \quad (7.19)$$

and the mismatch factor at the output of the first transistor is:

$$m_2 = \frac{a_2^2 g_{ie}}{g_{oe}} \quad (7.20)$$

Similarly, at the output of the second transistor (not shown in Fig. 7.4) we have

$$m_3 = \frac{a_3^2 g_{ie}}{g_{oe}} \quad (7.21)$$

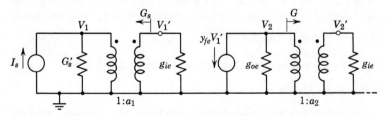

**Fig. 7.4.** First stage of the amplifier at the resonant frequency and with interaction neglected.

For any given stage, a small alignability factor can be obtained either by making $m_o$ large or $m_i$ small (see Eq. 7.18). However, if we assume that all interior stages are identical in design, then the mismatch factors for the interior stages will also be identical. Hence to make $k$ small, all of the mismatch factors must be either large or small. For typical transistors, $g_{ie}$ is much larger than $g_{oe}$; hence it is easier to make $m$ large rather than small, because smaller transformer ratios are required (see Eqs. 7.20 and 7.21). Thus the usual practice is to obtain a small alignability factor by providing large values of $m$ at each interstage.

The factor $m_1$ at the first transistor is often fixed by additional specifications imposed on the amplifier. For example, $G_s$ may be specified to provide optimum signal-to-noise performance by the amplifier, or $m_1$ may be specified to provide an impedance match between the amplifier and the source of signals. Thus designing multistage tuned amplifiers usually involves, among other things, choosing an alignability factor for each stage and then determining the required value of $m$ at the output of each transistor. The value of the transformation ratio required in each stage can then be determined by Eq. 7.20.

### 7.3.2 Power Gain

In deciding whether to neutralize or mismatch to obtain the desired alignability factor, it is necessary to estimate the power gain that can be obtained by each of these methods. The maximum available power gain for one completely unilateralized stage is given by Eq. 8.53 of ECP as:

$$A_p(\omega_0)\bigg|_{\substack{G=g_{oe}\\y_{re}=0}} = \frac{|y_{fe}|^2}{4g_{ie}g_{oe}} \tag{7.22}$$

where we assume that the neutralization is achieved by a capacitor, so that $g_{ie}$ and $g_{oe}$ are unchanged by the neutralization, and $y_{fe}$ is a composite parameter that includes the small effect of the neutralizing element. This is the gain that can be obtained by neutralizing perfectly and matching impedances at the output of the stage. Experience suggests that it also represents about the greatest gain that can be obtained reliably in a well-designed amplifier.

For designs based on mismatch, the gain $A_p(\omega_0)$ is, from ECP, Eq. 8.55,

$$A_p(\omega_0) = \frac{m_0|y_{fe}|^2}{g_{ie}g_{oe}(1 + m_o)^2 - \text{Re}[y_{fe}y_{re}(1 + m_o)]} \quad (7.23)$$

where $m_o$ is the mismatch factor at the output of the interior stage. Here $y_{fe}$ is the true forward transfer ratio of the transistor alone. If Eq. 7.23, indicates a large sacrifice in gain compared to Eq. 7.22, it becomes necessary to neutralize or to use a combination of neutralization and mismatching to obtain the desired alignability factor.

## 7.4 ILLUSTRATIVE DESIGN EXAMPLES

### 7.4.1 Design Using Mismatch to Achieve Alignability

The first example to be considered is the design of an amplifier using silicon epitaxial transistors to meet the following specifications:

Output must be delivered to a 100-ohm resistive load;
Apparent source resistance as seen by input transistor to be 500 ohms for good noise performance, when actual source resistance is 100 ohms
Power gain $A_p(\omega_0) = 10{,}000$ or 40 db
Center frequency $= 60$ mc or $3.8 \times 10^8$ rad/sec
Over-all half-power bandwidth $= 4$ mc

The admittance parameters for the transistor at 60 mc with $V_{CE} = 5$ volts and $I_C = 5$ ma are:

|          | $g$      | $C$     | $C'$    |
|----------|----------|---------|---------|
| $y_{ie}$ | 7 mmhos  | 16 pf   | 10 pf   |
| $y_{fe}$ | 33       | $-117$  | $-20$   |
| $y_{re}$ | 0        | $-2.1$  | $-2.1$  |
| $y_{oe}$ | 1        | 5.3     | 2.0     |

## Sec. 7.4 Illustrative Design Examples

To examine the stability of this transistor at 60 megacycles, following the method of Sec. 8.3.4 of ECP, we calculate:

$$g_{ie} - \frac{\text{Re}\,[y_{fe}y_{re}]}{2g_{oe}} - \frac{|y_{fe}y_{re}|}{2g_{oe}} = 7 + \frac{(44)(0.8)}{2} - \frac{(55)(0.8)}{2}$$

$$= 2.6 \text{ mmhos}$$

Because the result is greater than zero, the transistor itself is unconditionally stable when used in narrowband tuned amplifiers at this frequency.

To obtain an estimate of the minimum number of transistors that will provide the specified gain, we determine the maximum unilateralized gain from Eq. 7.22:

$$A_p(\omega_0)\Big|_{\substack{g_{oe}=G \\ y_{re}=0}} = \frac{|y_{fe}|^2}{4g_{ie}g_{oe}} = \frac{(55)^2}{(4)(7)(1)} = 108, \text{ or } 20.3 \text{ db}$$

(Note that we have here neglected the small effect ($\approx 2\%$) of the neutralizing element on $|y_{fe}|$.) Thus two neutralized stages can, it appears, supply the required 40 db of gain. However, if we allow one db for the loss in each coil, then two transistors will not suffice. On the other hand, three neutralized stages would provide substantially more gain than is required. We therefore examine the possibility of leaving the transistors unneutralized and mismatching to obtain good alignability. At this point, we choose a value of $k = 0.2$ for the alignability factor in each stage.

A simplified circuit diagram for the input circuit and the first stage of the amplifier is shown in Fig. 7.4. The tuned circuits on each side of the transistor are assumed to be in resonance, so that the reactive elements cancel out. Also, the small effects of the interaction have been neglected, in accordance with the discussion in Sec. 7.3. The mismatch factor at the input to the amplifier is fixed by the specification that $G_s = 2$ millimhos. Thus

$$m_1 = \frac{g_{ie}}{G_s} = \frac{7}{2} = 3.5$$

Then the value of $m_2$, the mismatch factor at the output of the first transistor, must be determined to provide the desired alignability factor:

$$k = \frac{|y_{fe}y_{re}|}{2g_{ie}g_{oe}(1 + 1/m_1)(1 + m_2)} = \frac{(0.8)(55)}{(2)(7)(1)(1.29)(1 + m_2)}$$

$$= \frac{2.43}{1 + m_2}$$

Thus, for $k = 0.2$, $m_2 = 11.2$. Accordingly, the gain of the first stage is, from Eq. 7.23,

$$A_{p1}(\omega_0) = \frac{(11.2)(55)^2}{(7)(91)(12.2)^2 + (44)(0.8)(12.2)} = 23.1$$

The second stage of the amplifier also has the configuration shown in Fig. 7.4, so the calculations proceed as for the first stage. The results are summarized in Table 7.1.

TABLE 7.1

|  | First Stage | Second Stage | Third Stage |
|---|---|---|---|
| $m$ | $m_1 = 3.5$<br>$m_2 = 11.2$ | $m_3 = 13.5$ | $m_4 = 13.5$ |
| Stage gain | 23.1<br>(13.6 db) | 20.7<br>(13.2 db) | 20.7<br>(13.2 db) |

The total gain for the three stages is thus:

$$A_p(\omega_0) = 13.6 + 13.2 + 13.2 = 40 \text{ db}$$

which is just equal to the power-gain specification (if we quite arbitrarily ignore coil losses).

The transformation ratio required at the input to the amplifier is:

$$a_1^2 = \frac{G_s'}{G_s} = \frac{10}{2} = 5$$

where $G_s$ and $G_s'$ are identified in Fig. 7.4. Similarly,

$$a_2^2 = \frac{G}{g_{ie}} = \frac{m_2 g_{oe}}{g_{ie}} = \frac{(11.2)(1)}{7} = 1.6$$

## Sec. 7.4 Illustrative Design Examples

The remaining transformation ratios are determined in a similar manner.

Now the tuned circuits must be designed to provide the specified center frequency and bandwidth. We assume that it will be possible to synchronously tune the amplifier. To determine the bandwidth that the individual stages must have in order to provide the specified over-all bandwidth, we find from Eq. 7.11, with $n = 4$ because we have four tuned circuits in the amplifier, that:

$$\text{Single stage bandwidth} = \frac{G_a}{C_a} = \frac{8\pi \times 10^6}{\sqrt{2^{1/4} - 1}}$$

$$= 57.8 \times 10^6 \text{ rad/sec or 9.2 mc}$$

The input circuit and the first stage of the amplifier are shown in Fig. 7.5. In this diagram, all of the parameters have been reflected across the transformers to the input and output terminals of the transistor and the small effects of the interaction have been neglected. Thus the tuning capacitance $C_2$ in Fig. 7.5 can be chosen to provide the desired bandwidth as follows:

$$C_a = \frac{g_{oe} + a_2{}^2 g_{ie}}{\text{bandwidth}} = \frac{(1 + m_2) g_{oe}}{\text{bandwidth}}$$

$$= \frac{12.2 \times 10^{-3}}{57.8 \times 10^6} = 211 \text{ pf}$$

Now, from Eq. 7.7:

$$C_2 = C_a - \frac{C_{oe} + C'_{oe}}{2} - a_2{}^2 \frac{(C_{ie} + C'_{ie})}{2}$$

$$= 211 - \frac{7.3}{2} - 1.6 \left(\frac{26}{2}\right) = 186 \text{ pf}$$

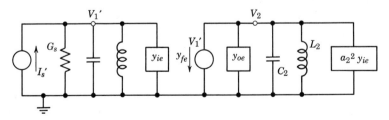

**Fig. 7.5.** First stage of the amplifier with interaction neglected.

The inductance required in this tuned circuit to provide the desired center frequency is:

$$L_2 = \frac{1}{\omega_0^2(C_2 + C_{oe} + a_2{}^2 C_{ie})}$$

$$= \frac{1}{(3.8)^2(10^{16})(186 + 5.3 + 25.6) \times 10^{-12}} = 0.032 \; \mu\text{h}$$

This value of inductance is rather small, and it would be desirable to use an additional impedance transformation as discussed in Chapter 8 of ECP to obtain a more convenient value.

The above calculations complete the design of one of the four tuned circuits in the amplifier. The parameter values for the remaining tuned circuits are determined in the same manner; hence, the calculations will not be carried out here.

### 7.4.2 Design Using Mismatch and Neutralization to Achieve Alignability

As a second example, let us consider a design in which we use both mismatch and neutralization to achieve the desired alignability. Assume that the amplifier must meet the following specifications:

Output to be delivered to a 1,000-ohm resistive load
Input impedance matched to a 1,000-ohm resistive source
Power gain $A_p(\omega_0) = 2 \times 10^6$, or 63 db
Center frequency = 480 kc, or $3 \times 10^6$ rad/sec
Over-all half-power bandwidth = 10 kc

The admittance parameters for the transistor at 480 kc with $V_{CE} = 5$ volts and $I_C = 2$ ma are:

|        | $g$      | $C$    | $C'$   |
|--------|----------|--------|--------|
| $y_{ie}$ | 1 mmho  | 67 pf  | 50 pf  |
| $y_{fe}$ | 50      | 0      | 0      |
| $y_{re}$ | 0       | $-10$  | $-10$  |
| $y_{oe}$ | 0.15    | 50     | 25     |

This transistor at 480 kc is potentially unstable (see ECP, Sec. 8.3.5).

## Sec. 7.4 Illustrative Design Examples

The maximum unilateralized gain for this transistor is, if we again neglect the small effect of the neutralizing element on $|y_{fe}|$:

$$A_p(\omega_0)\bigg|_{\substack{g_{oe}=G \\ y_{re}=0}} = \frac{|y_{fe}|^2}{4g_{ie}g_{oe}} = \frac{(50)^2}{(4)(1)(0.15)} = 4170, \text{ or } 36.2 \text{ db}$$

Thus two transistors in cascade are capable of supplying the required gain of 63 db with some margin to spare. We now check to see if the required gain can be obtained with the transistors unneutralized and with impedances mismatched to provide a suitable alignability factor. As in the preceding example, we choose $k = 0.2$ for each stage. The detailed calculations of $m$ and the stage gains are identical to those in Sec. 7.4.1. The results are summarized in Table 7.2.

**TABLE 7.2**

|  | Stage 1 | Stage 2 |
|---|---|---|
| $m$ | $m_1 = 1$ <br> $m_2 = 11.5$ | $m_3 = 22$ |
| Stage gain | 1230 <br> (31 db) | 690 <br> (28.4 db) |

Thus the gain of the two stages in cascade is 59 db, so if we allow 2 db for coil losses, we have 6 db less gain than required. Therefore, to meet the gain specification, we are faced with the same design alternatives listed in the first example. In this case, we will investigate what can be accomplished by using a combination of neutralization and mismatching.

If the loss from mismatch can be reduced by a factor of two in each stage with the aid of partial neutralization, then an additional 3 db of gain will be contributed by each stage, and the gain specification will be met. To increase the gain by a factor of two in the first stage, we require:

$$A_p(\omega_0) = \frac{m_2(50)^2}{(1)(0.15)(1+m_2)^2} = 1230 \times 2$$

or

$$m_2^2 - 4.8m_2 + 1 = 0$$

$$m_2 = 2.4 \pm 2.2$$

There are two mismatch factors that yield the desired gain, and they are reciprocals. Since we want $m_2$ large to provide a small alignability factor, we choose $m_2 = 4.6$.

We can now determine the magnitude of $y_{re}$ that is needed to obtain the desired value for $k$ from Eq. 7.18. Since $m_1 = 1$ in the first stage, and since $m_2 = 4.6$ yields the desired gain, we obtain, for $k = 0.2$,

$$|y_{re}| = \frac{(0.2)(2)(1)(0.15)(2)(5.6)}{50} = 0.013 \text{ mmho}$$

Thus, to achieve the desired result in the first stage, the neutralization must be effective enough to reduce $|y_{re}|$ by a factor of about 2 from its unneutralized value of 0.03 mmho.

To achieve a similar result in the second stage, we find using the technique shown above that $m_3 \approx 10$. Now, substituting into Eq. 7.18, we obtain, for $k = 0.2$,

$$|y_{re}| \cong 0.013 \text{ mmho}$$

If the neutralizing element is adjusted to neutralize completely the nominal value of the feedback admittance, $y_{re} = -j0.03$ millimho, then a production-line spread of $\pm j0.013$ millimho, or $\pm 43\%$, in this parameter can be tolerated without having $k$ exceed the design value of 0.2, and $k$ will be very small for units having $y_{re}$ close to the nominal value.

The remainder of the design problem consists of determining the transformation ratios and the parameter values for the tuned circuits. The computational procedures are the same as those presented in the first example at the beginning of this section, so they will not be repeated here.

**PROBLEMS**

**P7.1** A tuned amplifier using silicon planar transistors is to be designed to meet the following specifications:
    Output delivered to a 50-ohm resistive load
    Apparent source resistance seen by the first transistor to be 400 ohms for good noise performance
    Power gain $A_p(\omega_o) = 30$ db
    Center frequency = 45 mc
    Over-all half-power bandwidth = 3 mc

The transistor parameters in millimhos at 45 mc with $V_{CE} = 9$ volts and $I_C = 4$ ma are:

$$y_{ie} = 4 + j5 \qquad y_{re} = -j\,0.2$$
$$y_{fe} = 70 - j45 = 83\,\angle -33° \qquad y_{oe} = 0.1 + j\,0.4$$

Tuned circuits are to be used at the input, at the output, and in each interstage network. The transistors are to be left unneutralized, and impedances are to be mismatched to provide an alignability factor $k = 0.15$ in each stage.

(a) Find the maximum unilateralized gain for the transistor in db.
(b) Find the mismatch factors and the power gain $A_p(\omega_o)$ in db for each stage.
(c) Determine the number of transistors needed to meet the specifications.
(d) Specify the bandwidth in megacycles for each tuned circuit.

**P7.2** Complete the calculations of $m_3$, $m_4$, and the gains for the example in Sec. 7.4.1, and hence verify the numbers given in Table 7.1.

**P7.3** Calculate $m_1, m_2, m_3$, and the gain for each stage of the amplifier in Sec. 7.4.2, and hence verify the values given in Table 7.2.

# 8

# Interrelations Between Frequency Domain, Time Domain, and Circuit Parameters

## 8.0 INTRODUCTION*

This chapter is concerned with developing approximations that simplify calculations and give insight into the relation between sinusoidal frequency response, impulse and step response, circuit parameters, and algebraic expressions involving complex frequency. Some of these approximations have already been developed in Chapters 1 and 2, but in this chapter the approximations are explored in more detail and extended to a larger class of problems. Also, the emphasis here is on expressing the response in both the time and frequency domains.

In Chapter 1 it was seen that the sum of a set of circuit time constants ($\Sigma \tau_{jo}$), which could be found by inspection, was equal to the sum of the negative-reciprocal natural frequencies. This same sum of time constants, which we will call the first "time moment," $T_1$, was also seen to be closely related to frequency response, and was useful for approximate calculation of half-power bandwidth, $\omega_h \cong T_1^{-1}$. The approximations based on $T_1$ were used

* This chapter is somewhat more advanced than the preceding ones, in that it requires some previous experience with Fourier and Laplace transforms.

principally for analysis of amplifiers which had natural frequencies that were significantly different from one another because of interaction between various amplifier stages. In this chapter the approximation is extended by means of a second time moment, $T_2$, so as to be suitable for amplifiers with many closely spaced poles. The second time moment is equal to the square root of the sum of squares of the reciprocal natural frequencies, and can be calculated by inspection of the circuits. It will be seen that the two time moments can often lead to a good approximation for the impulse and step response for an important class of circuits. Moreover, they eliminate a difficult computational problem.

Before developing the approximations it is worthwhile to carry through a numerical example which illustrates the methods and problems of exact analysis. This same numerical example will also be used throughout the chapter as an illustration of the accuracy and simplicity of the approximate methods.

## 8.1 DIRECT CALCULATION OF FREQUENCY AND TIME RESPONSE

Let us first work a numerical example to illustrate the problem of calculating the frequency and time response for a known $A(s)$. Assume that $H(s) \equiv A(s)/A_0$ is given by:

$$H(s) = \frac{1}{(1 + 0.7s)(1 + 0.5s)(1 + 0.4s)(1 + 0.3s)(1 + 0.1s)} \tag{8.1a}$$

$$= \frac{1}{1 + 2s + 1.5s^2 + 0.52s^3 + 0.0809s^4 + 0.0042s^5} \tag{8.1b}$$

($s$ in units of $10^6$ sec$^{-1}$)

$H(s)$ is the normalized gain. It has a value of unity in the pass band which is assumed to extend down to dc. According to Eq. 8.1a, there are five zeros at $s = \infty$ and there are five natural frequencies located at:

$$s_1 = -(0.7)^{-1}, \quad s_2 = -(0.5)^{-1}, \quad s_3 = -(0.4)^{-1},$$

$$s_4 = -(0.3)^{-1}, \quad s_5 = -(0.1)^{-1}$$

($s$ in units of $10^6$ sec$^{-1}$)

For convenience of future discussion we will define the negative reciprocals of the natural frequencies as the "characteristic times" which will be denoted by $\tau_k$. Thus for this example we have:

$$\tau_1 = 0.7, \ \tau_2 = 0.5, \ \tau_3 = 0.4, \ \tau_4 = 0.3, \ \tau_5 = 0.1$$
($\tau$ in units of microseconds)

In Chapter 2 we saw various methods of approximating and plotting magnitude and phase response from either the factored or unfactored form of Eq. 8.1. By one of these methods we can arrive at the curves shown in Fig. 8.1. To find the half-power frequency, $\omega_h$, accurately we can expand either Eq. 8.1a or 8.1b in powers of $\omega^2$ as follows:

$$|H(j\omega)|^{-2} =$$

$$\begin{cases} (1 + 0.7^2\omega^2)(1 + 0.5^2\omega^2)(1 + 0.4^2\omega^2)(1 + 0.3^2\omega^2)(1 + 0.1^2\omega^2) \\ \qquad\qquad\qquad \text{and also} \qquad\qquad\qquad\qquad\qquad (8.2a) \\ (1 - 1.5\omega^2 + 0.0809\omega^4)^2 + \omega^2(2 - 0.52\omega^2 + 0.0042\omega^4)^2 \quad (8.2b) \end{cases}$$

$$= 1 + \omega^2 + 0.3318\omega^4 + 0.0445\omega^6$$
$$+ 0.00217681\omega^8 + 1.764 \times 10^{-5}\omega^{10} \quad (8.2c)$$

Note that the polynomial in Eq. 8.2c has all positive coefficients, and hence it is a monotonic function of $\omega^2$; it is evident from Eq. 8.2a that this will always be the case if all characteristic times are real. For small $\omega$ we can neglect all but the first few terms in Eq. 8.2. For example, the calculation of $\omega_h$ depends very little on higher order terms as we see by setting $|H(j\omega)|^{-2}$ equal to 2 in Eq. 8.2c:

Two terms: $1 + \omega_h{}^2 \cong 2$;

$$\omega_h \cong 1 \qquad\qquad\qquad \text{(see Eq. 1.16)}$$

Three terms: $1 + \omega_h{}^2 + 0.3318\omega_h{}^4 \cong 2$

$$\omega_h{}^{-2} \cong 0.5 + \sqrt{(0.5)^2 + .3318} \quad \text{or} \quad \omega_h \cong 0.890$$

All terms: $\quad \omega_h = 0.882 \qquad\qquad\qquad\qquad\qquad (8.3)$

Although the calculation of $\omega_h$ is not unduly complicated, it is desirable to find simpler approximate methods of computing $\omega_h$ accurately and directly from either the factored or unfactored form of Eq. 8.1 or, better yet, to compute $\omega_h$ simply by inspection of the

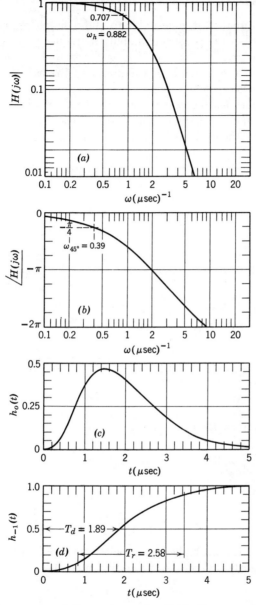

**Fig. 8.1.** Frequency and time response for illustrative example: $H^{-1}(s) = (1 + 0.7s)(1 + 0.5s)(1 + 0.4s)(1 + 0.3s)(1 + 0.1s)$.

network. Moreover, it is desirable to find a simple approximation which describes the phase and amplitude frequency response for $\omega$ significantly greater than $\omega_h$. The methods to be developed are extensions of the methods of Chapter 1 to the case of amplifiers with many closely spaced poles.

Before investigating approximate methods for describing $H(s)$, let us compute the impulse and step response for $H(s)$ given by Eq. 8.1. The theory is assumed to be familiar in general, but the method will be reviewed briefly.

The first step is to expand $H(s)$ into partial fractions as follows:

$$H(s) = \frac{1}{(1 + 0.7s)(1 + 0.5s)(1 + 0.4s)(1 + 0.3s)(1 + 0.1s)} \tag{8.4}$$

$$= \frac{A_1}{1 + 0.7s} + \frac{A_2}{1 + 0.5s} + \frac{A_3}{1 + 0.4s} + \frac{A_4}{1 + 0.3s} + \frac{A_5}{1 + 0.1s} \tag{8.5}$$

where:

$$A_1 = [(1 + 0.7s)H(s)]_{(s = -1/0.7)}$$

$$= \frac{0.7^4}{(0.7 - 0.5)(0.7 - 0.4)(0.7 - 0.3)(0.7 - 0.1)}$$

$$= \frac{343}{14.4} \times 0.7$$

$$A_2 = [(1 + 0.5s)H(s)]_{(s = -1/0.5)} \tag{8.6}$$

$$= -\frac{1125}{14.4} \times 0.5$$

And similarly,

$$A_3 = \frac{1024}{14.4} \times 0.4, \quad A_4 = -\frac{243}{14.4} \times 0.3, \quad A_5 = \frac{1}{14.4} \times 0.1$$

If there had been higher order poles, the expansion would have taken a slightly different form. For example, if:

$$H(s) = \frac{P_1(s)}{Q_1(s)} = \frac{P_1(s)}{(1 + \tau s)^m Q_2(s)};$$

$$\text{(with } P_1(s), Q_2(s) \neq 0 \text{ for } \tau s = -1) \tag{8.7}$$

### Sec. 8.1 Direct Calculation of Frequency and Time Response

a partial fraction expansion yields

$$H(s) = \frac{A_m}{(1+\tau s)^m} + \frac{A_{m-1}}{(1+\tau s)^{m-1}} + \cdots \frac{A_{m-k}}{(1+\tau s)^{m-k}}$$

$$+ \cdots \frac{A_1}{1+\tau s} + \frac{P_2(s)}{Q_2(s)} \quad (8.8a)$$

$$A_m = \{(1+\tau s)^m H(s)\}_{s=-1/\tau}$$

$$A_{m-1} = \frac{1}{\tau}\left\{\frac{d}{ds}[(1+\tau s)^m H(s)]\right\}_{s=-1/\tau} \quad (8.8b)$$

$$\cdots$$

$$A_k = \frac{1}{\tau^k k!}\left\{\frac{d^k}{ds^k}[(1+\tau s)^m H(s)]\right\}_{s=-1/\tau}$$

The coefficient $A_m$ must be nonzero because there is an $m$th order pole at $s = -1/\tau$, but the other coefficients may be zero. $P_2(s)$ is a uniquely determined polynomial, and to complete the partial fraction expansion, $P_2(s)/Q_2(s)$ is expanded in a similar manner with up to $m$ terms for each $m$th order pole.

The impulse response, $h_o(t)$, can be found from Eq. 8.5 (or 8.8 if higher order poles are present) by means of the standard transform:

$$\frac{A}{(1+\tau s)^m} \leftrightarrow u_{-1}(t)\,\frac{A}{\tau}\,\frac{1}{(m-1)!}\left(\frac{t}{\tau}\right)^{m-1} e^{-t/\tau} \quad (8.9)$$

where $H(s) \leftrightarrow h_o(t)$ denotes a transform pair, and $u_{-1}(t)$ = unit step occurring at $t = 0$. Applying Eq. 8.9 to Eq. 8.5 we have:

$h_0(t) = u_{-1}(t)$

$$\times \left[\frac{A_1}{0.7}e^{-t/.7} + \frac{A_2}{0.5}e^{-t/.5} + \frac{A_3}{0.4}e^{-t/.4} + \frac{A_2}{0.3}e^{-t/.3} + \frac{A_1}{0.1}e^{-t/.1}\right]$$

$$= u_1(t)\,\frac{1}{14.4}\,[343e^{-t/0.7} - 1125e^{-t/0.5} + 1024e^{-t/0.4} \quad (8.10a)$$

$$- 243e^{-t/0.3} + e^{-t/0.1}]$$
($t$ in microseconds)

The step response is given by:

$$h_{-1}(t) = \int_0^t h_o(t)\,dt = u_{-1}(t)[A_1(1 - e^{-t/0.7}) + A_2(1 - e^{-t/0.5}) \ldots]$$

$$= u_{-1}(t)\frac{1}{144}[-2401e^{-t/0.7} + 5625e^{-t/0.5}$$
$$- 4096e^{-t/0.4} + 729e^{-t/0.3} - e^{-t/0.1} + 144] \quad (8.10b)$$
($t$ in microseconds)

Equations 8.10 are plotted in Fig. 8.1c and d.

Two useful parameters for describing the step response, $h_{-1}(t)$, are the "delay" and "rise time." For convenience of measurement, it is customary to define a delay, $T_d$, as the time elapsed between the application of the input step and the instant that the output reaches 50% of its final value. Similarly, for measurement convenience, the rise time, $T_r$, is customarily defined as the time required for the output to increase from 10% to 90% of its final value. These definitions are indicated in Fig. 8.2 along with the commonly used definitions of "overshoot" and "undershoot." We see from Fig. 8.1 that for our numerical example, there is no overshoot or undershoot, and delay and rise times are:

$$T_d = 1.89 \ \mu\text{sec} \qquad T_r = 2.58 \ \mu\text{sec}$$

The parameters $T_d$ and $T_r$ in the time domain are closely related to $\omega_{45°}$ ($\equiv$ frequency for $\angle H(j\omega) = \pm 45°$) and $\omega_h$ in the frequency domain, as will be demonstrated in the next section.

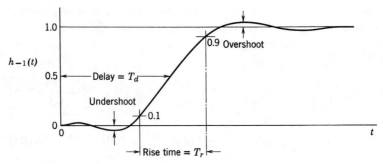

**Fig. 8.2.** Definition of terms which are commonly used to describe the step response.

In performing calculations from Eqs. 8.10, we must keep in mind that high accuracy is needed because we are dealing with small differences between comparable large terms. For example, typical values are:

$$h_o(t=0) = \frac{1}{14.4}[343 - 1125 + 1024 - 243 + 1] = 0$$

$$h_o(t=0.42) = \frac{1}{14.4}[188.24 - 485.67 + 358.34 - 59.92 + 0.01]$$

$$= \frac{1.00}{14.4} = 0.068$$

$$h_o(t=2.1) = \frac{1}{14.4}[17.08 - 16.87 + 5.38 - 0.22 + 0]$$

$$= \frac{5.37}{14.4} = 0.32$$

In short, the calculations require about two more significant figures than are valid in the answer. If there had been more poles, or if the original poles had been closer together, the accuracy problem would have been even more severe. This computational difficulty emphasizes the need for approximate methods which lead to impulse and step responses which can be easily calculated.

## 8.2 APPROXIMATIONS FOR H(jω) AND h(t)

### 8.2.1 *Time moments*

The example worked out in the last section gives a fair idea of the methods and problems of direct calculation of $H(j\omega)$ and $h_o(t)$. The objective of this section is to develop approximations which can lead to quick estimates of either $H(j\omega)$ or $h_0(t)$, as well as to give insight into relations between circuit parameters, frequency response, and time response. *The primary emphasis is on finding approximations for the special case of an amplifier with no peaking in the magnitude versus frequency plot and no overshoot in the step response.*

The key idea is to represent a response by certain parameters which have dimensions of time and are *directly* and *simply* related to frequency response, step response, and circuit elements. We have called these parameters "time moments," for reasons which will become apparent. Consider first an $H(s)$ with all its zeros at infinity so that:

$$H^{-1}(s) = (1 + \tau_1 s)(1 + \tau_2 s) \cdots (1 + \tau_k s) \cdots (1 + \tau_n s) \quad (8.11)$$

For a physical system, the characteristic times, $\tau_k$, in Eq. 8.11 are real or occur in conjugate pairs. We will define the time moments $T_1$, $T_2$, etc. for Eq. 8.11 as follows:

$$T_1 \equiv \sum_{k=1}^{n} \tau_k \ ; \quad T_2{}^2 \equiv \sum_{k=1}^{n} \tau_k{}^2 \ ; \quad \cdots \quad T_m{}^m \equiv \sum_{k=1}^{n} \tau_k{}^m \quad (8.12)$$

In words, $T_1$ is the sum of the characteristic times, $T_2$ is the square root of the sum of the squares of the characteristic times, etc. We are concerned primarily with the use of $T_1$ and $T_2$ as parameters for describing the response.

There are several possible advantages of expressing a response in terms of $T_m$ instead of in terms of $\tau_k$. One important reason is that the time moments can be calculated *without* factoring a polynomial. Assume, for example, that $H^{-1}(s)$ is given as a polynomial in $s$:

$$H^{-1}(s) = 1 + a_1 s + a_2 s^2 + \cdots \quad (8.13)$$

We saw in Chapter 1 that $T_1 = a_1$ (i.e., Eq. 1.20 with $s_k{}^{-1} = -\tau_k$). To find higher-order time moments, Eq. 8.11 is multiplied out into a polynomial in $s$ and the coefficients of $s$ are equated to the $a_k$ coefficients of Eq. 8.13. For example, $T_2$ is found as follows:

$$a_1 = \tau_1 + \tau_2 + \cdots \tau_n = \sum \tau_k = T_1$$

$$a_2 = \tau_1\tau_2 + \tau_1\tau_3 + \cdots + \tau_2\tau_3 + \cdots + \tau_{n-1}\tau_n \quad (8.14)$$

$$= \sum_{j=1}^{n} \sum_{k=1}^{j-1} \tau_k\tau_j \equiv \sum^{k<j} \tau_k\tau_j = \tfrac{1}{2}[(\sum \tau_k)^2 - \sum \tau_k{}^2] = \tfrac{1}{2}[T_1{}^2 - T_2{}^2]$$

Thus,

$$T_2{}^2 = a_1{}^2 - 2a_2 \quad (8.15)$$

Higher-order moments can be found in a similar manner with $T_m{}^m$ expressed in terms of the first $m + 1$ terms of Eq. 8.13 (e.g., $T_3{}^3 = a_1{}^3 - 3a_1 a_2 + 3a_3$, and $T_4{}^4 = a_1{}^4 - 4a_1{}^2 a_2 + 2a_2{}^2 + 4a_1 a_3 - 4a_4$). The important conclusion is that $T_1$ and $T_2$ can be calculated from the first three terms of $H^{-1}(s)$ and thus they can be found without factoring a high-order algebraic equation.

Another important property of $T_1$ and $T_2$ is the fact that they can be calculated directly from the circuit. In particular, for a network containing only capacitive energy storage, we see by reference to Chapter 1, Eq. 1.39 (with $a_o = 1$), that $T_1 = a_1 = \sum \tau_{jo}$ where $\tau_{jo}$ is the time constant of the $j$th capacitor with all other capacitors open-circuited. In other words, for this important special case of no inductors, $T_1$ is *both* the sum of the characteristic times *and also* the sum of the open-circuit time constants. In Sec. 8.2.4 we shall find a similar relation between $T_2{}^2$ and a set of new circuit time constants. First, however, let us see how $T_1$ and $T_2$ can be used to find a simple approximation to the frequency and time response.

### 8.2.2 *The Gaussian approximation*

If we take the logarithm of Eq. 8.11 and expand about $s = 0$ we have:

$$\begin{aligned}\ln H(s) &= -\ln(1 + \tau_1 s) - \ln(1 + \tau_2 s) - \ldots - \ln(1 + \tau_n s) \\ &= -(\tau_1 + \tau_2 + \ldots + \tau_n)s + \tfrac{1}{2}(\tau_1{}^2 + \tau_2{}^2 + \ldots + \tau_n{}^2)s^2 \\ &\quad - \tfrac{1}{3}(\tau_1{}^3 + \tau_2{}^3 + \ldots + \tau_n{}^3)s^3 \ldots \quad (8.16) \\ &= -T_1 s + \tfrac{1}{2}T_2{}^2 s^2 - \tfrac{1}{3}T_3{}^3 s^3 + \ldots\end{aligned}$$

(provided $|\tau_k s| < 1$ for all $k$)

Note that in Eq. 8.16, if $s = j\omega$, then the odd terms (i.e., $T_1 s$, $T_3{}^3 s^3/3$, etc.) will be pure imaginary while the even terms will be pure real. This is true even if the $\tau_k$ are complex, because complex $\tau_k$ always come in conjugate pairs. For example, $(\tau_r + j\tau_i)^2 + (\tau_r - j\tau_i)^2 = 2\tau_r{}^2 - 2\tau_i{}^2$ which is a real (but not necessarily positive) number. Hence, we have:

$$\angle H(j\omega) = -T_1 \omega + \tfrac{1}{3}(T_3 \omega)^3 - \ldots \quad (8.17a)$$

$$\ln |H(j\omega)| = -\tfrac{1}{2}(T_2 \omega)^2 + \tfrac{1}{4}(T_4 \omega)^4 - \ldots \quad (8.17b)$$

For small enough $\omega$, we can neglect all but the first terms in Eqs. 8.17. Assume for a moment that $T_1, T_2 > 0$, and let us neglect terms of higher order than $\omega^2$ (the conditions under which we can neglect higher-order terms will be considered later). On dropping terms higher than $\omega^2$, Eqs. 8.17 become

$$\angle H(j\omega) \cong -T_1\omega \qquad (8.18a)$$

$$|H(j\omega)| \cong e^{-(1/2)(T_2\omega)^2} \qquad (8.18b)$$

Equations 8.18 are plotted in Fig. 8.3. The right hand side of Eq. 8.18a is called a "linear" phase response because $\angle H(j\omega)$ is proportional to $\omega$. The right hand side of Eq. 8.18b is called a "Gaussian" response because it has the $e^{-x^2}$ form. Note that when Eqs. 8.18 are plotted on a logarithmic frequency scale, as in Fig. 8.3, both the linear phase and Gaussian magnitude fall off slowly at first, then more and more rapidly as $\omega$ increases.

Solutions of a number of numerical examples suggest that if there is no peaking in the frequency response and no overshoot in the step response, then Eq. 8.18a is valid up to at least $\omega = \omega_{45°}$ and Eq. 8.18b is valid up to at least $\omega = \omega_h$. Thus it is worthwhile to calculate $\omega_{45°}$ and $\omega_h$ on the basis of Eqs. 8.18. The result is as follows:

$$\angle H(j\omega_{45°}) = -\omega_{45°}T_1 = -\pi/4$$

Thus,

$$\omega_{45°} = \pi/4T_1 = 0.7854 T_1^{-1} \qquad (8.19a)$$

Similarly:

$$|H(j\omega_h)|^2 = e^{-(T_2\omega_h)^2} = \tfrac{1}{2}$$

Thus,

$$\omega_h = \sqrt{\ln 2}\; T_2^{-1} = 0.8326 T_2^{-1} \qquad (8.19b)$$

At this point there may appear to be a dilemma as to whether $\omega_h$ depends on $T_1$ or $T_2$. In Chapter 1 we saw that $\omega_h \cong T_1^{-1}$ and now we see that, if the Gaussian approximation is valid, $\omega_h \cong 0.833 T_2^{-1}$. Remember, however, that Chapter 1 considered the special case where interaction between stages caused significant splitting of the natural frequencies. In such cases the values of $T_1$ and $T_2$ can not be very different, so either $T_1$ or $T_2$ can be used to estimate $\omega_h$. For example, if successive characteristic times differ by at least a factor of two, it can be shown that $T_1/T_2$ can never exceed $\sqrt{3}$, regardless of how many poles there may be. Thus, in this case, either $T_1$ or $T_2$

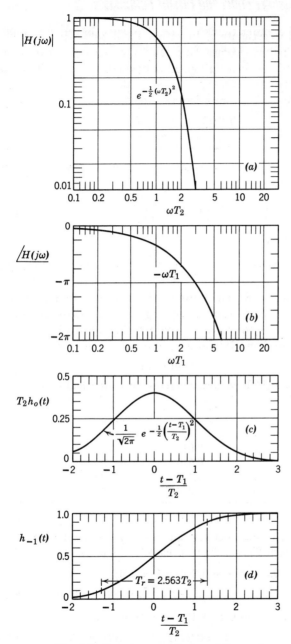

**Fig. 8.3.** Gaussian magnitude and linear phase.

can be used to estimate $\omega_h$ to within about 30%. However, if the number of poles becomes large and the spacing between poles become small, the calculation of $\omega_h$ by means of $T_1$ will be quite inaccurate, but the Gaussian approximation becomes quite good. For example, if there are $n$ identical poles we have:

$$H^{-1}(s) = (1 + \tau s)^n \qquad (8.20a)$$

$$\omega_h = (\sqrt{2^{1/n} - 1})\,\tau^{-1} \qquad (8.20b)$$

But $2^{1/n} - 1$ can be expanded in a power series in $1/n$ about $1/n = 0$. Thus,

$$2^{1/n} - 1 = \frac{1}{n}\ln 2 + \frac{1}{2!}\left(\frac{1}{n}\ln 2\right)^2 + \ldots \qquad (8.21)$$

and for large $n$:

$$\omega_h = \sqrt{2^{1/n} - 1}\,\tau^{-1} \cong \left(\sqrt{\frac{\ln 2}{n}}\right)\tau^{-1} = \sqrt{\ln 2}\,T_2^{-1} \qquad (8.22)$$

On comparing Eq. 8.22 with Eq. 8.19b, we see that for large $n$, $\omega_h$ for the $n$ identical-pole example converges to the Gaussian value for $\omega_h$. Moreover, the convergence is very fast and Eq. 8.19b is never in error by more than 17%, as is indicated by Table 8.1.

**TABLE 8.1**
**Bandwidth reduction factor for $n$ identical poles**

| $n$ | $\sqrt{2^{1/n} - 1}$ | $\sqrt{\ln 2/n}$ |
|---|---|---|
| 1 | 1 | .833 |
| 2 | 0.643 | .589 |
| 3 | 0.510 | .480 |
| 4 | 0.435 | .416 |
| 5 | 0.386 | .373 |

As an example of bandwidth calculation, consider the five-pole example of Eq. 8.1. Let us call $\omega_{h1}$ and $\omega_{h2}$ the bandwidths as calculated from $T_1$ and $T_2$, respectively. Thus we have:

$$T_1 = 0.7 + 0.5 + 0.4 + 0.3 + 0.1 = 2.0;$$

$$\omega_{h1} = T_1^{-1} = 0.500 \times 10^6 \text{ sec}^{-1} \qquad (8.23a)$$

$$T_2^2 = 0.7^2 + 0.5^2 + 0.4^2 + 0.3^2 + 0.1^2 = (1.0)^2;$$

$$\omega_{h2} = 0.833 T_2^{-1} = 0.833 \times 10^6 \text{ sec}^{-1} \qquad (8.23b)$$

The correct value is $\omega_h = 0.882$ so $\omega_{h2}$ is in error by less than 6% while $\omega_{h1}$ is in error by about 43%. The $\omega_{h2}$ approximation is better because there are five poles spaced fairly close together.

One advantage of the Gaussian approximation is that it allows us to estimate the impulse and step response by means of standard tabulated functions. To find the time domain response we use the transform pairs*

$$e^{-j\omega T_1} H(j\omega) \leftrightarrow h_o(t - T_1) \tag{8.24a}$$

$$A e^{-(1/2)(T_2\omega)^2} \leftrightarrow \frac{A}{T_2\sqrt{2\pi}} e^{-(1/2)(t/T_2)^2} \tag{8.24b}$$

Thus Eqs. 8.18 transform into:

$$h_o(t) = \frac{1}{T_2} \frac{1}{\sqrt{2\pi}} \exp\left[-1/2 \left(\frac{t - T_1}{T_2}\right)^2\right] \tag{8.25a}$$

$$h_{-1}(t) = \int_{-\infty}^{t} h_o(t)\, dt = 1/2 \left[\operatorname{erf}\left(\frac{t - T_1}{\sqrt{2}\, T_2}\right) + 1\right] \tag{8.25b}$$

Both $e^{-x^2}$ and $\operatorname{erf}(x)$ (the "error function" of $x$) are tabulated functions. We thus find $h_o(t)$ and $h_{-1}(t)$ are as shown in Fig. 8.3. Note that strictly speaking Eqs. 8.25 describe an unrealizable response because there is a finite output before the input is applied. However, for $T_1 > 2T_2$, $h_{-1}(t = 0)$ is less than 2.3% of the final value and the Gaussian approximation is very close to a realizable response.

It is particularly convenient to compute the delay and rise times (as defined in Fig. 8.2) from Eqs. 8.25. From the tabulated functions we find:

$$T_d = T_1 \tag{8.26a}$$

$$T_r = 2.563 T_2 \tag{8.26b}$$

We conclude from Eqs. 8.26 that the delay depends only on $T_1$ while the rise time depends only on $T_2$. This fact is not surprising since, according to Eqs. 8.18, low-frequency phase shift, which is related to delay, depends only on $T_1$ while bandwidth, which is related to rise time, depends on $T_2$.

* Equations 8.24 are written in terms of $\omega$ in order to avoid convergence problems associated with the use of $s$, because $h_o(t) \neq 0$ for $t < 0$.

An important result can be deduced by multiplying Eq. 8.19b by Eq. 8.26b:

$$\omega_h T_r = \frac{0.8326}{T_2} \times 2.563 T_2 = 2.134 \quad (8.27)$$

or

$$f_h T_r = \frac{2.134}{2\pi} = 0.3396 \quad (8.28)$$

Similarly,

$$f_{45°} T_1 = \frac{\pi/4}{2\pi} = 1/8 = 0.1250 \quad (8.29)$$

The fact that for a Gaussian response $T_r f_h = 0.34$ has important consequences in the design of amplifiers. It is apparent from Fourier series considerations that bandwidth and rise time are related, and Eq. 8.28 is merely a quantitative expression of this fact for amplifiers which have a Gaussian-like response. For this class of amplifiers, the fact that $T_r$ is proportional to $T_2$ gives us a clue as to how the rise time of a cascade of several noninteracting amplifiers can be computed from the rise times of the individual amplifiers. Suppose, for example, that we have two amplifiers, designated $a$ and $b$, and that each one is unilateral and has a high input and a low output impedance. Suppose, moreover, that we calculate or measure $T_2$ for each amplifier when excited from a low impedance source and connected to a high impedance load. The assumed input and output impedances and unilateral behavior imply that the characteristic times of the individual amplifiers are unchanged by the cascading. Thus $T_2$ for the cascade, $T_{2c}$, is given by

$$T_{2c}^2 = \sum \tau_{ka}^2 + \sum \tau_{kb}^2 = T_{2a}^2 + T_{2b}^2 \quad (8.30)$$

where $T_{2a}$ and $T_{2b}$ are $T_2$ for individual amplifiers. Since, according to Eq. 8.26b, $T_r$ for the cascade depends on $T_2$ for the cascade, we see from Eq. 8.30 that "rise times of cascaded, noninteracting amplifiers with Gaussian-like responses add quadratically." That is,

$$T_{rc}^2 = T_{ra}^2 + T_{rb}^2 \quad (8.31)$$

One example of the use of this result occurs in the experimental measurement of pulse rise time. If a pulse generator produces a pulse with a leading edge resembling $h_{-1}(t)$ in Fig. 8.3b, and if a noninteracting amplifier and oscilloscope are cascaded to observe this pulse, then the rise time of the observed pulse will be approximately the square root of the sum of the squares of the separate rise times of the original pulse, the amplifier, and the oscilloscope. Note, however, that this calculation depends on the Gaussian-like rise times of the pulse generator, amplifier, and oscilloscope *and* it is essential that the three elements do not interact. Note specifically that rise times of individual stages of a transistor amplifier do *not* add quadratically because of the interaction of one stage with the others.

### 8.2.3 Bounds on the response with negative-real poles and infinite zeros

For the important special case of all zeros at infinity and all poles on the $\omega = 0$ axis, the Gaussian response and the single time-constant response form bounds on the realizable response. We will now derive these bounds, which are indicated in Fig. 8.4 and in Eq. 8.32:

$$e^{-(1/2)(\omega T_2)^2} < |H(j\omega)| \le \left|\frac{1}{1 + j\omega T_2}\right| \qquad (8.32)$$

The proof of Eq. 8.32 can be demonstrated from an expansion of $|H(j\omega)|^{-2}$ in powers of $\omega^2$:

$$|H(j\omega)|^{-2} = (1 + \tau_1^2\omega^2)(1 + \tau_2^2\omega^2)\ldots(1 + \tau_k^2\omega^2)\ldots(1 + \tau_n^2\omega^2) \qquad (8.33a)$$

$$= 1 + \omega^2(\textstyle\sum\tau_k^2) + \omega^4(\textstyle\sum\tau_k^2\tau_j^2) + \ldots + \omega^{2m}(\overbrace{\textstyle\sum\tau_k^2\tau_j^2\ldots}^{m\ \tau^2\text{-factors}})$$

$$\ldots + \omega^{2n}(\tau_1\tau_2\ldots\tau_n); \quad k < j < \ldots \le n \qquad (8.33b)$$

$$\equiv 1 + c_1\omega^2 + c_2\omega^4 + \ldots + c_n\omega^{2n} \qquad (8.33c)$$

Since we are assuming that $\tau_k$ is always positive real, it is clear that all terms in Eq. 8.33b are positive. Hence one bound on $|H(j\omega)|$ is:

$$|H(j\omega)|^{-2} \ge 1 + \omega^2(\textstyle\sum\tau_k^2) = 1 + (\omega T_2)^2$$

or

$$|H(j\omega)| \le \frac{1}{|1 + j\omega T_2|} \qquad (8.34)$$

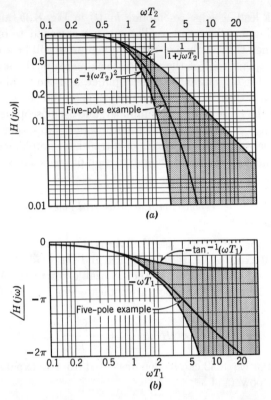

**Fig. 8.4.** Bounds on frequency response for $H(s)$ with all zeros at infinity and all characteristic times positive real.

Proof of the other bound requires a little more work. First, note that a bound on the coefficient of $\omega_4$ (i.e., $c_2$) can be derived as follows:

$$c_2 = \sum_{}^{k<j} \tau_k^2 \tau_j^2 = \tfrac{1}{2}[(\textstyle\sum \tau_k^2)^2 - (\textstyle\sum \tau_k^4)] < \tfrac{1}{2}(\textstyle\sum \tau_k^2)^2 = \tfrac{1}{2} T_2^4 \quad (8.35a)$$

Similarly, there is a bound on the coefficient $c_3$:

$$c_3 = \sum_{}^{k<j<i} \tau_k^2 \tau_j^2 \tau_i^2 = \frac{1}{3!}[(\textstyle\sum \tau_k^2)^3 - (\textstyle\sum \tau_k^6 + 2\sum_{}^{k<j} \tau_k^2 \tau_j^4)] < \frac{1}{3!} T_2^6$$
(8.35b)

and in general:

$$c_m = \frac{1}{m!}[T_2^{2m} - (T_{2m}^{2m} + \ldots)] < \frac{1}{m!} T_2^{2m} \quad (8.35c)$$

On combining Eqs. 8.35 and 8.33, the bound on $|H(j\omega)|^{-2}$ is:

$$|H(j\omega)|^{-2} < 1 + (\omega T_2)^2 + \frac{1}{2!}(\omega T_2)^4 + \frac{1}{3!}(\omega T_2)^6 + \ldots = e^{(\omega T_2)^2}$$

$$(\omega T_2)^2 < \infty$$

or

$$|H(j\omega)| > e^{-(1/2)(\omega T_2)^2} \qquad (8.35d)$$

By reference to Eq. 8.35c, we see that for $c_m$ to approach the upper bound it is certainly necessary that $T_2^{2m} \gg T_{2m}^{2m}$ for $m > 1$. This, in turn, requires that no one or two time constants can be dominant and, in fact, there must be many almost equal values of $\tau_k$. To illustrate this fact, assume that $\tau_k = \tau$ for all $k$. We then have:

$$|H(j\omega)|^{-2} = 1 + n(\tau\omega)^2 + \frac{(n)(n-1)}{2}(\tau\omega)^4 + \ldots$$

and for large $n$, since $n\tau^2 = T_2^2$ and $n(n-1) \cong n^2$, etc., $|H(j\omega)|^{-2}$ is approximately equal to, but less than, the quantity

$$1 + (T_2\omega)^2 + \frac{1}{2!}(T_2\omega)^4 + \ldots = e^{(\omega T_2)^2}$$

which is consistent with Eq. 8.35d.

In summary, Eq. 8.32 specifies that if we plot $|H(s)|$ as a function of normalized frequency, $\omega T_2$, the response must lie between a single pole response and the Gaussian response. The Gaussian response, in turn, is essentially equivalent to the response for an infinitely large number of identical characteristic times.

The proof of Eq. 8.32 can be extended to prove the corresponding phase bounds:

$$-(\omega T_1) < \angle H(j\omega) \leq -\tan^{-1}(\omega T_1) \qquad (8.36)$$

These two bounds are shown in Fig. 8.4 along with the plot of the five-pole example from Eq. 8.1. It is clear that the response for the example does, in fact, lie within the bounds. It is also clear from Fig. 8.4 that the bounds diverge drastically for $\omega > 2\omega_h$, but the interesting conclusion is that the half power frequency, $\omega_h$, and the frequency where the phase shift is 45°, $\omega_{45°}$, *always* lie within the range:

$$0.833 < \omega_h T_2 \leq 1.0 \qquad (8.37a)$$
$$0.785 < \omega_{45°} T_1 \leq 1.0 \qquad (8.37b)$$

Hence, the linear phase and Gaussian magnitude approximations are almost always valid for computing $\omega_{45°}$, $\omega_h$, $T_d$, and $T_r$ for a system with all real poles and all zeros at infinity.

We have seen in Eq. 8.28 and 8.29 that the Gaussian response implies $f_h T_r = 0.340$ and $f_{45°} T_d = 0.125$. It is interesting to note that the other bound in Eq. 8.32 leads to a similar result. For a single time constant amplifier we have,

$$T_d = 0.6832 T_1; \quad f_{45°} = (2\pi T_1)^{-1}; \quad T_d f_{45°} = 0.1087 \quad (8.38a)$$

$$T_r = 2.197 T_2; \quad f_h = (2\pi T_2)^{-1}; \quad T_r f_h = 0.3498 \quad (8.38b)$$

The numerical values of $T_d f_{45°}$ and $T_r f_h$ in Eqs. 8.38 are quite close to the values for the Gaussian approximation in Eqs. 8.29 and 8.28, and experience seems to indicate that the single time constant response and Gaussian response represent, qualitatively, the extreme possibilities for an impulse or step response when there are only negative real poles and infinite zeros. Moreover, experimental evidence indicates that if the frequency response has no peaking and the step response has no overshoot, then the choice of pole and zero locations does not have a dominant effect on the major features of the response and, for example, most linear amplifiers with negligible peaking or overshoot have $T_r f_h \cong 0.35$.

### 8.2.4 *Relations between $T_2{}^2$ and circuit time constants*

In order to calculate $T_2$ from circuit parameters, we could use the relation $T_2{}^2 = a_1{}^2 - 2a_2$ (Eq. 8.15) and calculate $a_1$ and $a_2$ separately. However, this procedure is not usually the easiest: calculation of $a_2$ requires one term for every pair of capacitors, and there are accuracy problems because $2a_2$ may be nearly equal to $a_1{}^2$. Moreover, this approach does not lead to a simple relation between circuit parameters and $T_2{}^2$. Thus we prefer to derive an expression for $T_2{}^2$ directly in terms of a set of circuit time constants similar to the expression $T_1 = \sum \tau_{j_0}$ as derived in Chapter 1.

A proof very similar to the one in Chapter 1, Sec. 1.2.2 leads to the result, analogous to Eq. 1.39 with $a_0 = 1$, that

$$a_2 = \sum^{i<j} (r_{ii} r_{jj} - r_{ij} r_{ji}) C_i C_j \quad (8.39)$$

where $r_{ij}$ is the $ij$ element in the $R$ matrix (i.e., the inverse $G$ matrix). In words, $r_{ii}$ is the dc resistance seen by the $i$th capacitor,

## Sec. 8.2 Approximations for $H(j\omega)$ and $h(t)$

and $r_{ij}$ the dc voltage across the $i$th capacitor, caused by a unit dc current source in parallel with the $j$th capacitor. Since $r_{jj}C_j = \tau_{jo}$ (as defined in Chapter 1, Eq. 1.39ff) we can write Eq. 8.39 in the form

$$a_2 = \sum_{}^{i<j} \tau_{io}\tau_{jo} - \sum_{}^{i<j} r_{ij}r_{ji}C_iC_j$$

$$= \tfrac{1}{2}[(\sum \tau_{jo})^2 - \sum \tau_{jo}^2] - \sum_{}^{i<j} r_{ij}r_{ji}C_iC_j \qquad (8.40)$$

By means of Eq. 8.15, $T_2^2$ can be expressed in terms of $\tau_{jo}$ etc.:

$$T_2^2 = a_1^2 - 2a_2 = T_1^2 - [(\sum \tau_{jo})^2 - \sum \tau_{jo}^2] + 2\sum_{}^{i<j} r_{ij}r_{ji}C_iC_j$$

$$= \sum \tau_{jo}^2 + 2\sum_{}^{i<j} r_{ij}r_{ji}C_iC_j \qquad (8.41)$$

By eliminating the constraint on the summation indices $i$ and $j$, Eq. 8.41 can be written in the form

$$T_2^2 = \sum \tau_{ij}^2 \quad \text{(summed over all } i \text{ and } j\text{)} \qquad (8.42)$$

where $\tau_{ij}^2 \equiv r_{ij}r_{ji}C_iC_j$.

Equation 8.42 is quite analogous to the expression for $T_1$ in Eq. 1.39 and simply states that $T_2^2$ is the sum of the squares of a set of circuit time constants which can be calculated by considering all pairs of capacitors and their interactions. In a transistor amplifier frequently all of the $\tau_{ij}^2$ are positive so the calculation tends to be accurate and one can often identify particular pairs of capacitors as contributing the dominant terms. Moreover, the computation can be simplified because many of the $\tau_{ij}$ are negligible (i.e., the $r_{ij}r_{ji}$ product is often very small) and because approximations often allow us to estimate $\tau_{ij}$ directly from the $\tau_{jo}$ time constants.

As an illustration of how $T_2^2$ can be calculated directly from a circuit, consider the example of Chapter 1, Fig. 1.3$b$, page 7. Let us define $C_{\pi 1} = C_1$, $C_{\mu 1} = C_2$, $C_{\pi 2} = C_3$, etc., and define $R_{10}$, $R_{L1}$, etc., as in Ch. 1. Figure 1.3$b$ can then be drawn in the form of Fig. 8.5. The terms $\tau_{11}^2$, $\tau_{22}^2$ etc., are calculated very easily from Eqs. 1.45, 1.47, and 1.49 to 1.52 because $\tau_{11}^2 = \tau_{10}^2$, $\tau_{22}^2 = \tau_{20}^2$, etc. Thus:

$$\tau_{11}^2 = (R_{10}C_1)^2; \qquad \tau_{22}^2 = (R_{10} + R_{L1} + g_m R_{10} R_{L1})^2 C_2^2 \qquad (8.43a)$$

$$\tau_{33}^2 = (R_{20}C_3)^2; \qquad \tau_{44}^2 = (R_{20} + R_{L2} + g_m R_{20} R_{L2})^2 C_4^2 \qquad (8.43b)$$

$$\tau_{55}^2 = (R_{30}C_5)^2; \qquad \tau_{66}^2 = (R_{30} + R_{L3} + g_m R_{30} R_{L3})^2 C_6^2 \qquad (8.43c)$$

The only new problem is to calculate the $\tau_{ij}$ caused by interaction of two capacitors.

Consider first the interaction of $C_1$ and $C_2$. If $C_2$ is replaced by a unit dc current source, the dc voltage across $C_1$ will be $r_{12}$, given by

$$r_{12} = R_{10}$$

And, similarly, if $C_1$ is replaced by a unit dc current source, the dc voltage across $C_2$ will be $r_{21}$, given by

$$r_{21} = R_{10}(1 + g_m R_{L1})$$

and thus

$$\tau_{12}^2 = R_{10}^2(1 + g_m R_{L1})C_1 C_2 \qquad (8.44a)$$

The signs of $r_{12}$ and $r_{21}$ will depend on the choice of reference polarity for the voltage across $C_1$ and $C_2$, but as long as a specific choice is adhered to, the product $r_{12}r_{21}$ will not depend on the sign convention. For the circuit of Fig. 8.5 all of the $\tau_{ij}^2$ are positive so there is no sign difficulty.

The interactions between $C_3$ and $C_4$ and between $C_5$ and $C_6$ are found in a similar way with results analogous to Eq. 8.44a:

$$\tau_{34}^2 = R_{20}^2(1 + g_m R_{L2})C_3 C_4 \qquad (8.44b)$$

$$\tau_{56}^2 = R_{30}^2(1 + g_m R_{L3})C_5 C_6 \qquad (8.44c)$$

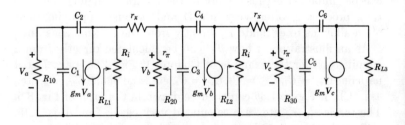

$R_{10} = 80$, $R_i = 780$, $r_x = 50$, $r_\pi = 400$, $R_{L3} = 50$, $g_m = 0.2$,
$C_1 = C_3 = C_5 = 78.5$, $C_2 = C_4 = C_6 = 2.5$
$(R_{L1} = R_{L2} = 285, R_{20} = R_{30} = 270)$

**Fig. 8.5.** Three stage amplifier based on Fig. 1.3b (Values in ohms and picofarads).

Next, consider the interaction between $C_2$ and $C_3$. From Fig. 8.5 we have:

$$r_{23} = \frac{R_i}{R_i + r_x} R_{20}; \qquad r_{32} = (1 + g_m R_{10}) R_{L1} \frac{r_\pi}{r_\pi + r_x}$$

$$\tau_{23}^2 = (1 + g_m R_{10}) R_{L1} R_{20} \frac{R_i r_\pi}{(R_i + r_x)(r_\pi + r_x)} C_2 C_3 \qquad (8.45a)$$

and similarly,

$$\tau_{45}^2 = (1 + g_m R_{20}) R_{L2} R_{30} \frac{R_i r_\pi}{(R_i + r_x)(r_\pi + r_x)} C_4 C_5 \qquad (8.45b)$$

Finally, consider the interaction between $C_2$ and $C_4$ and between $C_4$ and $C_6$:

$$r_{24} = r_{23}; \qquad r_{42} = r_{32}(1 + g_m R_{L2}) \qquad (8.46a)$$

$$\tau_{24}^2 = (1 + g_m R_{10}) R_{L1} (1 + g_m R_{L2}) R_{20} \frac{R_i r_\pi}{(R_i + r_x)(r_\pi + r_x)} C_2 C_4$$

and similarly,

$$\tau_{46}^2 = (1 + g_m R_{20}) R_{L2} (1 + g_m R_{L3}) R_{30} \frac{R_i r_\pi}{(R_i + r_x)(r_\pi + r_x)} C_4 C_6$$
$$(8.46b)$$

All other $\tau_{ij}^2$ are zero. For example, a dc current source in parallel with $C_3$, $C_4$, $C_5$, or $C_6$ could not affect the dc voltage across $C_1$. If we put in numerical values in the above expressions for $\tau_{ij}^2$, the resulting interaction time constants are as given in the matrix in Table 8.2. Thus, $\tau_{11}^2 = .39 \times 10^{-16}$, $\tau_{12}^2 = \tau_{21}^2 = .73 \times 10^{-16}$, etc.

### TABLE 8.2

$\tau_{ij}^2$ interaction time constants for circuit of Fig. 8.5.
Unit of $\tau^2$ is $(10^{-8} \text{ sec})^2 = 10^{-16} \text{ sec}^2$.

|  |  |  |  |  |  | Row Sum |
|---|---|---|---|---|---|---|
| .39 | .73 |  |  |  |  | 1.12 |
| .73 | 1.52 | 2.14 | 3.97 |  |  | 8.36 |
|  | 2.14 | 4.49 | 8.30 |  |  | 14.93 |
|  | 3.97 | 8.30 | 15.90 | 6.95 | 2.44 | 37.56 |
|  |  |  | 6.95 | 4.49 | 1.58 | 13.02 |
|  |  |  | 2.44 | 1.58 | .56 | 4.58 |

$[\tau_{ij}^2] = $ above matrix

$T_2^2 = 79.57$

One of the chief advantages of finding $T_2^2$ by tabulating $\tau_{ij}^2$ as in Table 8.2, is that one can now identify the extent to which each capacitor contributes to the over-all bandwidth. For example, it is clear from Table 8.2 that $C_4$ has the dominant effect. If we set $C_4 = 0$, the value of $T_2^2$ can be calculated by eliminating the fourth row and column. The result is $T_2^2(C_4 = 0) = 20.35$ or practically a 2:1 reduction in $T_2$. Thus we could double the bandwidth by eliminating $C_4 (= C_{\mu 2})$. Clearly it is not possible to completely eliminate $C_{\mu 2}$, but at least we might consider taking steps to use a transistor with a low value of $C_\mu$ for the second stage.

We could also extend Table 8.2 to investigate the effect of additional stages. Table 8.3 shows the additional terms that would appear if a fourth stage were added. The added terms would

### TABLE 8.3

$\tau_{ij}^2$ for circuit of Fig. 8.5 when extended to a four-stage amplifier. Units of $10^{-16}$ sec$^2$.

$$[\tau_{ij}^2] = \begin{bmatrix} .39 & .73 & & & & & \\ .73 & 1.52 & 2.14 & 3.97 & & & \\ & 2.14 & 4.49 & 8.30 & & & \\ & 3.97 & 8.30 & 15.90 & 6.95 & 12.82 & \\ & & & 6.95 & 4.44 & 8.30 & \\ & & & 12.82 & 8.30 & 15.90 & 6.95 & 2.44 \\ & & & & & 6.95 & 4.49 & 1.58 \\ & & & & & 2.44 & 1.58 & .56 \end{bmatrix}$$

Added terms ⟶

$\Sigma = 66.53$

increase $T_2^2$ by $66.5 \times 10^{-16}$ sec$^2$ or, in other words, the value of $T_2^2$ would be almost doubled by going to four stages. As a matter of fact, $T_2^2$ increases by $66.5 \times 10^{-16}$ sec$^2$ for each additional stage after the third. No change in source or load could affect this value (because each stage is unilateral at low frequencies) so, in a sense, the numbers which most accurately describe the single stage in this iterated array are:

$$T_1 = 6.12 \times 10^{-8} \text{ sec} \qquad \text{Per stage}$$
$$T_2^2 = 66.5 \times 10^{-16} \text{ sec}^2 \qquad \text{values}$$

## Sec. 8.2 Approximations for $H(j\omega)$ and $h(t)$

There is an additional simplification that can be used to greatly facilitate the calculation of Table 8.2. This simplification takes advantage of three approximations:

(1) If the stage voltage gain is large, then $g_m R_{L1} \gg 1$ and hence,

$$R_{L1}(1 + g_m R_{10}) \cong R_{10} + g_m R_{10} R_{L1} + R_{L1} \quad (8.47)$$

and, similarly, for the second and third stage.

(2) If the stage current gain is large, then $g_m R_{10} \gg 1$ and hence,

$$R_{10}(1 + g_m R_{L1}) \cong R_{10} + g_m R_{10} R_{L1} + R_{L1} \quad (8.48)$$

and, similarly, for the second and third stage.

(3). If $r_x$ is not too large, then

$$\frac{R_i r_\pi}{(R_i + r_x)(r_\pi + r_x)} = 1 \quad (\text{i.e., } R_i \| r_\pi \gg r_x)$$

If these approximations are valid then life is very simple as far as $T_2^2$ calculations are concerned. Note, for example, that (see Eqs. 8.43a and 8.44a)

$$R_{10} C_1 R_{10}(1 + g_m R_{L1}) C_2 = \tau_{12}^2 \cong \tau_{11}\tau_{22}; \quad \tau_{34}^2 \cong \tau_{33}\tau_{44}; \quad \tau_{56}^2 \cong \tau_{55}\tau_{66} \quad (8.49)$$

and similarly,

$$\tau_{23}^2 \cong \tau_{22}\tau_{33}; \quad \tau_{45}^2 \cong \tau_{44}\tau_{55}; \quad \tau_{24}^2 \cong \tau_{22}\tau_{44}; \quad \tau_{46}^2 \cong \tau_{44}\tau_{66}$$

In short, the expression for $T_2^2$ becomes

$$T_2^2 = \sum \tau_{ii}\tau_{jj}\delta_{ij} \quad (8.50)$$

where $\delta_{ij} \cong 1$ for two interacting capacitors (i.e., $r_{ij}r_{ji} \neq 0$)

$\delta_{ij} \cong 0$ for two noninteracting capacitors (i.e., $r_{ij}r_{ji} = 0$).

The only critical problem is to recognize which capacitors interact. If, in Fig. 8.5, we let $r_x = 0$ then the interacting capacitors are those which share a common node.

According to Eq. 8.50 we can calculate an approximate form of Table 8.2 by simply cross-multiplying the appropriate values of the open circuit time constants calculated in Chapter 1. The result is shown in Table 8.4. The entries in this table are found by simply multiplying two open-circuit time constants. For example, $\tau_{11} = 0.63 \times 10^{-8}$ and $\tau_{22} = 1.23 \times 10^{-8}$ so $\tau_{12}^2 \cong 0.63 \times 1.23 = 0.77 \times 10^{-16}\ \text{sec}^2$. Note that Table 8.4 does, in fact, form a good approximation to Table 8.2 and that the resulting value of $T_2$ is in error by less than 5%. The approximation in Eq. 8.50 is, of course, limited to structures such as the one in Fig. 8.5, and is valid only if the $g_m$ term dominates the impedance seen by $C_\mu$. Thus Eq. 8.50 is not valid if there is a low value resistor in parallel with $C_\mu$.

### TABLE 8.4

Approximate calculation of $\tau_{ij}^2$ from Eq. 8.50. Entry $\tau_{ij}^2$ is calculated by cross-multiplying the numbers at left of row $i$ and at top of column $j$. Values of $\tau_{11} = 0.63$, $\tau_{22} = 1.23$, etc., are from Eqs. 1.45 to 1.52. Units are $10^{-16}\ \text{sec}^2$.

|  | (0.63) | (1.23) | (2.12) | (4.00) | (2.12) | (0.75) | $\sum$ |
|---|---|---|---|---|---|---|---|
| (0.63) | 0.40 | 0.77 |  |  |  |  | 1.17 |
| (1.23) | 0.77 | 1.51 | 2.61 | 4.92 |  |  | 9.81 |
| (2.12) |  | 2.61 | 4.49 | 8.48 |  |  | 15.58 |
| (4.00) |  | 4.92 | 8.48 | 16.00 | 8.48 | 3.00 | 40.88 |
| (2.12) |  |  |  | 8.48 | 4.49 | 1.59 | 14.56 |
| (0.75) |  |  |  | 3.00 | 1.59 | .56 | 5.15 |

$$T_2'^2 = 87.15$$

In summary, the value of $T_2{}^2$ can be computed directly from a circuit with good accuracy and the calculations are often not much more difficult than the calculation of the open-circuit time constants. The calculations of $T_2{}^2$ by means of $\tau_{ij}^2$ has the advantage of identifying clearly the extent to which each capacitor affects the value of $T_2{}^2$.

### 8.2.5 Time moments when zeros are present

The preceding examples have all been based on the assumption that $H(s)$ had all of its zeros at $s = \infty$. Many practical examples have finite zeros so it is worthwhile to extend the definition of $T_1$ and $T_2$ in an appropriate manner. The proper definitions of $T_1$ and $T_2$ are clear if we expand $\ln H(s)$ as before. Let

$$H(s) = \frac{(1 + \tau_{z1}s)(1 + \tau_{z2}s) \ldots (1 + \tau_{zk}s) \ldots}{(1 + \tau_1 s)(1 + \tau_2 s) \ldots (1 + \tau_k s) \ldots} \quad (8.51)$$

$\tau_k \equiv$ characteristic times; $\quad \tau_{zk} \equiv$ characteristic zero-times

## Sec. 8.2 Approximations for $H(j\omega)$ and $h(t)$

Then $\ln H(s) = 1 - (\sum \tau_k - \sum \tau_{zk})s + \frac{1}{2}(\sum \tau_k^2 - \sum \tau_{zk}^2)s^2 \ldots$

(provided $|\tau_k s|$ and $|\tau_{zk} s| < 1$ for all $k$) (8.52)

By analogy with Eq. 8.16, the time moments can be defined as

$$T_m^m = \sum \tau_k^m - \sum \tau_{zk}^m \qquad (8.53)$$

The presence of zeros as well as poles will invalidate the bounds given in Eq. 8.32, but the time moments are still very useful for estimating response. As an example, suppose we would like to estimate the increase in frequency response caused by inductive shunt peaking similar to that described in Chapter 5. For simplicity, assume that there are four identical, noninteracting stages with each stage described by the model in Fig. 8.6 (with $r_\pi = \infty$). The values of $C$ and $R$ are presumed to be fixed by the transistor characteristics and the desired gain. The shunt peaking inductance, $L$, is assumed to be adjusted to the largest possible value which does not give overshoot in the step response (i.e., all real poles). Straightforward analysis leads to a per-stage gain, $A_{v1}$, given by:

$$A_{1v} = g_m R \frac{1 + s(L/R)}{1 + s(RC) + s^2(LC)} \equiv A_0 \frac{(1 + \tau_3 s)}{(1 + \tau_1 s)(1 + \tau_2 s)} \qquad (8.54)$$

$$\tau_1 + \tau_2 = RC \qquad \text{and} \qquad \tau_1 \tau_2 = LC$$

Moreover, as we wish to choose $L$ as large as possible for $\tau_1$ and $\tau_2$ real, we must have

$$\tau_1 = \tau_2 = \frac{RC}{2} \quad \text{and} \quad \tau_1 \tau_2 = LC = \left(\frac{RC}{2}\right)^2; \quad L = \frac{R^2 C}{4}$$

$$A_{v1} = A_0 \frac{1 + (\tau s/4)}{[1 + (\tau s/2)]^2}; \quad \tau = \frac{RC}{2} \qquad (8.55a)$$

$$T_2^2 \text{ (for four stages)} = 4\left[\left(\frac{\tau}{2}\right)^2 + \left(\frac{\tau}{2}\right)^2 - \left(\frac{\tau}{4}\right)^2\right] = \frac{7}{4}\tau^2 \qquad (8.55b)$$

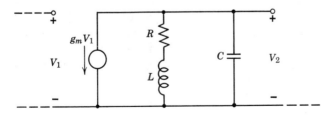

**Fig. 8.6.** Model of one stage of shunt peaked amplifier.

From Eq. 8.55b we see that for four stages the value of $T_2^2$ is $(7/4)\tau^2$ so we would estimate the half-power bandwidth as

$$\omega_h\left(L = \frac{R^2C}{4}\right) = \frac{0.833}{\sqrt{7/4}}\tau^{-1} = 0.630\tau^{-1} \qquad (8.56a)$$

Without shunt peaking, the bandwidth can be determined from Table 8.1:

$$\omega_h(L = 0) = 0.435\tau^{-1} \qquad (8.56b)$$

Thus the shunt peaking has increased the bandwidth by the factor $0.630/0.435 = 1.45$, or a 45% increase. For comparison with Eq. 8.56a, the true bandwidth with $L$ present is $\omega_h = 0.647\tau^{-1}$ so the approximation is accurate to better than 4%.

In using Eq. 8.53 we should appreciate that if the contribution of the zeros is large, then $|H(j\omega)|$ will not be a monotonically decreasing function and the Gaussian approximation is not reasonable. In the preceding example, note that the zero contributed only about 14% to $\tau_2^2$, so the Gaussian approximation is reasonable.

### 8.2.6 Time Domain Interpretation of $T_1$ and $T_2$

The preceding calculations show the importance of $T_1$ and $T_2$ in determining phase shift and bandwidth, but it remains to establish a specific relation between $T_1$, $T_2$ and the impulse response for the case where $|H(j\omega)|$ is not a simple function.

Probably the best way to relate $T_1$ and $T_2$ to the impulse response is in terms of a graphical center-of-gravity and radius-of-gyration interpretation. Figure 8.7 shows the relations which we will now prove. In order to make the proof as clear as possible, we will state the result in advance: "*The center of gravity of the area bounded by the impulse response occurs at $t = T_1$ and the radius of gyration of this area about its center of gravity is $T_2$.*" It should be evident from Fig. 8.7 that for a nonoscillatory impulse response there is a close connection between $T_d$ and $T_1$ and between $T_r$ and $T_2$. If there is significant oscillation in the impulse response, then, of course, the radius of gyration could even be negative, and the connection between $T_2$ and $T_r$ would become very tenuous. Thus, even though the theorem is true for all stable circuits, the usefulness is limited primarily to responses with negligible oscillations in the impulse response.

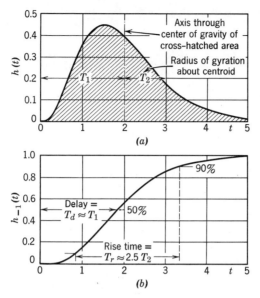

**Fig. 8.7.** Geometrical interpretation of first and second time moments, $T_1$ and $T_2$, and approximate relation to delay and rise times, $T_d$ and $T_r$.

In order to prove the geometrical time moment interpretation, it is convenient to use the following well known transform identities:

$$h_0(t) \leftrightarrow H(s) \qquad (8.57a)$$

$$th_0(t) \leftrightarrow -\frac{d}{ds}[H(s)] \qquad (8.57b)$$

$$h_0(t+\tau) \leftrightarrow H(s)e^{\tau s} \qquad \leftrightarrow \text{defines a transform pair} \qquad (8.57c)$$

$$\int_{-\infty}^{t} h_o(t)\, dt \leftrightarrow \frac{1}{s} H(s) \qquad (8.57d)$$

$$\lim_{t \to \infty} [h_0(t)] = \lim_{s \to 0} [sH(s)] \qquad (8.57e)$$

In order to find the center of gravity of the impulse response, we can use the following definition of the center of gravity:

$$\int_{-\infty}^{\infty} th_o(t+\tau)\, dt = 0 \text{ for } \tau = T_{cg}$$

where $T_{cg} \equiv$ value of $t$ at center of gravity of $h(t)$ (8.58)

**272** *Frequency, Time, and Circuit Parameters*

Using the identities in Eqs. 8.57, we can write Eq. 8.58 in the $s$-plane form:

$$\left\{\frac{d}{ds}[H(s)e^{\tau s}]\right\}_{s=0} = 0 \quad \text{for } \tau = T_{cg} \quad (8.59)$$

Assume now that $H(s)$ can be written as a ratio of polynomials normalized to unity for $s = 0$:

$$H(s) \equiv \frac{P(s)}{Q(s)} \equiv \frac{1 + b_1 s + b_2 s^2 + \cdots}{1 + a_1 s + a_2 s^2 + \cdots} \quad (8.60)$$

Combining Eqs. 8.59 and 8.60, we have:

$$\frac{d}{ds}[H(s)e^{\tau s}]_{s=0} = H(s)\left[\frac{P'(s)}{P(s)} - \frac{Q'(s)}{Q(s)} + T_{cg}\right]_{s=0} = 0 \quad (8.61)$$

where

$$P'(s) \equiv \frac{dP(s)}{ds}, \qquad Q'(s) \equiv \frac{dQ(s)}{ds} \quad (8.62)$$

To complete the proof we combine Eq. 8.60 and 8.61 and solve for $T_{cg}$, with the result:

$$T_{cg} = a_1 - b_1 = T_1 \quad (8.63)$$

To prove the time domain interpretation of $T_2$, we can use the following definition of the radius of gyration, $T_{rg}$, with respect to $t = T_{cg} = T_1$

$$T_{rg}^2 \equiv \frac{\int_{-\infty}^{\infty} t^2 h_o(t + T_1)\, dt}{\int_{-\infty}^{\infty} h_o(t + T_1)\, dt} = \left[\frac{(d^2/ds^2)[H(s)e^{T_1 s}]}{H(s)e^{T_1 s}}\right]_{s=0}$$

$$= \left[\frac{P''(s)P(s) - [P'(s)]^2}{P(s)} - \frac{Q''(s)Q(s) - [Q'(s)]^2}{Q(s)}\right]_{s=0}$$

$$= [(b_1^2 - 2b_2) - (a_1^2 - 2a_2)] = T_2^2 \quad (8.64)$$

In summary, we can interpret the impulse response geometrically as indicated in Fig. 8.7. This completes the objective of relating $T_1$ and $T_2$ to (1) poles and zeros, (2) magnitude and phase, (3) the impulse and step response, and (4) circuit parameters. Since $T_1$

and $T_2$ alone describe a good many of the properties of a system with small overshoot, they can often replace a more complicated description. Two systems with drastically different pole-zero configurations but with the same value of $T_1$ and $T_2$ and negligible overshoot can have very nearly the same response. As with most approximations, however, $T_1$ and $T_2$ do not describe the fine-structure of the response and thus, for example, they cannot predict the transient response near $t = 0$ or the frequency response for $\omega \gg \omega_h$. If there is significant overshoot in the step response, then $T_1$ and $T_2$ have much less significance as descriptive parameters.

## 8.2.7 The Lumped-Pole Approximation

If only two parameters are needed to characterize a response, it is apparent that we should be able to find one or more approximations which enable us to predict the response without tedious calculations. The Gaussian response is one such approximation, but this response violates the realizability condition because it predicts $h(t) \neq 0$ for $t < 0$. A much better approximation in this respect is to describe an amplifier by a "lumped-pole" approximation. The idea is to find a multiple-order pole which has the same value of $T_1$ and $T_2$ as the system being approximated. Thus we have:

$$H(s) \cong H_a(s) \equiv \frac{1}{(1 + \tau s)^n}; \quad T_1 = n\tau, \quad T_2^2 = n\tau^2 \quad (8.65a)$$

or

$$n = T_1^2/T_2^2, \quad \tau = T_2^2/T_1 \quad (8.65b)$$

From given values of $T_1$ and $T_2$, we can use Eqs. 8.65b to calculate $n$ and $\tau$. As an example of the accuracy of this approximation, consider the original five-pole example. If

$$H(s)^{-1} = (1 + 0.7s)(1 + 0.5s)(1 + 0.4s)(1 + 0.3s)(1 + 0.1s)$$

Then

$$T_1 = 2.0; \quad T_2^2 = 1.0$$

$$n = \frac{T_1^2}{T_2^2} = 4.0; \quad \tau = \frac{T_2^2}{T_1} = 0.5$$

and thus

$$H_a^{-1}(s) = (1 + \tau s)^n = (1 + 0.5s)^4 \quad (8.66)$$

This approximation can be compared with the exact response by comparing Figs. 8.4 and 8.8. The approximation predicts $|H(j\omega)|$ with a maximum error of about 16% up to $15f_h$, and at this frequency the gain is less than $10^{-3}$ times the midband value. In the time domain, the approximate step response is never in error by more than 0.01 (i.e., 1% of the final value) for all $t$. Hence, the lumped-pole approximation has a much wider range of validity than the Gaussian approximation. Note, however, that $n$ for the lumped-pole approximation is not equal to the number of poles in the original $H(s)$. The five-pole example required only a fourth-order lumped pole and, in general, $n$ need not even be an integer. For noninteger $n$, the impulse response can still be determined from Eq. 8.9 if $(m-1)!$ is replaced by $\Gamma(m)$ (i.e., the "gamma function" of $m$, which is a tabulated function).

Figure 8.9 shows how bandwidth and rise time are related to $n$, $T_1$, and $T_2$. By means of the plots in this figure, the parameters $f_h$ and $T_r$ can be easily determined without involved calculation. In the case of the five-pole example we have $n=4$ and $T_2 = 1$ μsec so, using Fig. 8.9, we find

$$f_h T_2 = 0.1384 \quad \text{or} \quad f_h = 138.4 \text{ kc} \qquad (8.67a)$$

$$\frac{T_r}{T_2} = 2.467 \quad \text{or} \quad T_r = 2.467 \text{ μsec} \qquad (8.67b)$$

The correct values are $f_h = 881.9/2\pi = 140.4$ kc and $T_r = 2.435$ μsec so the values in Eqs. 8.67 are accurate to better than 2%.

From Fig. 8.9 we see that, over a wide range of $n$,

$$f_h = \frac{0.14}{T_2} \; ; \quad T_r = 2.5 T_2 \qquad (8.68)$$

and thus:

$$f_h T_r = 0.35$$

These results are, of course, essentially the Gaussian approximations modified slightly to be a better approximation for moderate values of $n$.

### 8.3 LOW-FREQUENCY CONSIDERATIONS

The previous section has been concerned with high-frequency characteristics of "low-pass" amplifiers which "pass" all fre-

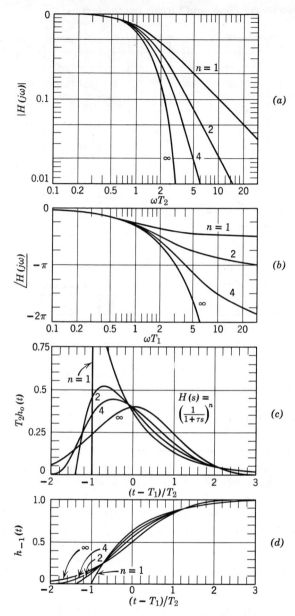

**Fig. 8.8.** Response of low-pass amplifier with $n$th order pole and all zeros at $s = \infty$.

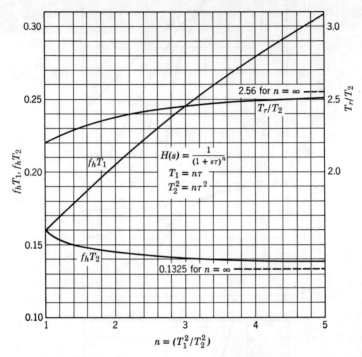

**Fig. 8.9.** Bandwidth and rise time for the lumped-pole approximation.

quencies up to about $\omega_h$. Fortunately, many of the approximations developed for these can be applied to the analysis of the low-frequency properties of ac-coupled amplifiers. If we neglect high-frequency effects, the ac-coupled amplifier becomes a "high-pass" amplifier which passes all frequencies above about its lower half-power frequency $\omega_l$.

In this section we are concerned with the properties of high-pass amplifiers having $H(s)$ of the form:

$$H(s) = \frac{s^n}{s^n + a_1 s^{n-1} + \ldots} = \frac{1}{1 + a_1 s^{-1} + a_2 s^{-2} + \ldots + a_n s^{-n}} \quad (8.69)$$

Evidently the change of variable $s \to s^{-1}$ converts Eq. 8.69 into a form identical with that of the low-pass amplifier described by Eq. 8.13.

To apply the results of the preceding section we can define "frequency moments," $\Omega_m$, analogous to the time moments discussed previously. The lowest one for the response given by Eq. 8.69 is:

$$\Omega_1 = a_1 = \sum \tau_k^{-1} \qquad (8.70)$$

The frequency moments can also be expressed directly in terms of a set of time constants, in this case *short-circuit* time constants. Based on Chapter 1 we have:

$$\Omega_1 = \sum g_{jj}/C_j \qquad (8.71)$$

where $g_{jj}$ is the $jj$ element of a *conductance* matrix; that is, $g_{jj}$ is the dc conductance facing $C_j$ when all other capacitors in the network are *short-circuited*, and the sum is taken over all capacitors in the circuit.

For calculations of step and frequency response down to, say, $0.1\omega_l$, we can use a lumped-pole approximation; the magnitude and phase can be deduced from Fig. 8.8, and step response curves are shown in Fig. 8.10.

The analogies between high-pass and low-pass amplifiers allow one to simplify circuit calculations. But one important difference between low-pass and high-pass amplifiers is that, in the latter,

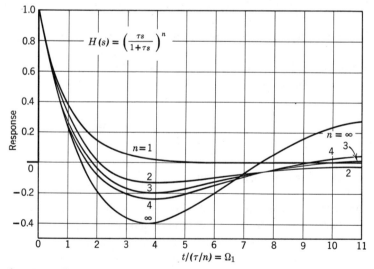

**Fig. 8.10.** Step response of high-pass amplifier with $n$th order pole and all zeros at $s = 0$.

## 278   Frequency, Time, and Circuit Parameters

phase-shift characteristics may dominate the response even when there is no feedback stability problem. To see why, consider the square-wave response of an amplifier with a single characteristic time $\tau$. The response will have the form shown in Fig. 8.11, which defines the "sag" (sometimes called "droop"), usually expressed as a percentage of the peak value. The sag should ideally be very small for good fidelity. However, a small sag requires a value of $f_l$ much less than the frequency of the square wave, $f_{sw}$, as can be seen by considering a numerical example.

Assume that we would like to have less than 1% sag for a 500 cps square wave, after it has passed through a high-pass amplifier with a single pole. When the sag is small, the exponential decay in Fig. 8.11 can be approximated by a linear decay. Thus,

$$\text{per cent sag} = 100(1 - e^{-T_{sw}/2\tau}) \cong 100\left(\frac{T_{sw}}{2\tau}\right) \tag{8.72}$$

For a 1% sag, $\tau = 50 T_{sw}$. So for a 500 cps square wave, $\tau = 0.1$ sec and $f_l = 1.6$ cps. In short, the value of $f_l$ must be more than 300 times smaller than the frequency of the square wave. But the magnitude of the gain of the amplifier for $f = 300 f_l$ is exceedingly close to its asymptotic value. So we surely cannot attribute the 1% sag to the magnitude imperfection.

The connection of sag with phase shift is exhibited by expressing the step response for small $t$ in terms of the frequency moment, $\Omega_1$. For small $t$, the step response, $h_{-1}(t)$, is given by the transform of $s^{-1}H(s)$ for large $s$. Thus, if

$$s^{-1}H(s) = \frac{s^{-1}}{1 + \Omega_1 s^{-1} + \ldots} \cong \frac{1}{s + \Omega_1} \tag{8.73}$$

then for small $t$,

$$h_{-1}(t) \cong u_{-1}(t) e^{-\Omega_1 t} \tag{8.74}$$

Hence $\Omega_1$ determines the sag in the square wave response for $f_{sw} \gg f_l$.

One method of combatting sag is to design a circuit with a zero whose characteristic zero-time is chosen to make $\Omega_1 = 0$ (in analogy with Eq. 8.53). The disadvantage is usually "peaking" in $|H(j\omega)|$; i.e., the "good" square wave response for $f_{sw} \gg f_l$ is obtained at the expense of a poor sinusoidal response for $f$ near $f_l$.

Another unique low-frequency problem is evident from a study of Fig. 8.10. Note that the step response is plotted as a function of

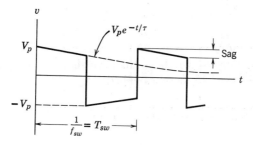

**Fig. 8.11.** Square wave response for single-pole, high-pass amplifier.

$nt/\tau = \Omega_1 t$. Thus the curves all correspond to amplifiers which have about the same amount of sag in the square wave response for $f_{sw} \gg f_t$. Observe, however, that as the number of poles increases, the response for large $t$ becomes increasingly large, and is oscillatory in nature. This "residual" response, for $t > \Omega^{-1}$, occurs after the important effect of the step is over, and is usually undesirable. It is particularly so for pulse amplifiers, because the residual response of one large pulse becomes confused with the initial response of succeeding smaller pulses.

Figure 8.10 shows that the residual response is greatly reduced if $n = 1$. Hence, it is common to design amplifiers with a single dominant characteristic time, or even to use a dc amplifier, just to avoid the residual response problem.

## PROBLEMS

**P8.1** Figure 8.12a shows a lumped model for a representation of base resistance at high frequencies. For circuit analysis it is desirable to represent this distributed problem by the lumped model shown in Fig. 8.12b where $r_x$ is to be chosen so that the input admittance, $y_i$, and the forward transadmittance, $y_f$, are as accurately described as possible. The analytic expression for $y_f$ (assuming large $n$ in Fig. 8.12a) is

$$y_f = g_m \frac{\tanh \sqrt{s\tau}}{\sqrt{s\tau}}; \qquad \tau = r_b C_\pi$$

$$= g_m \frac{1 + \frac{1}{3!}(s\tau) + \frac{1}{5!}(s\tau)^2 + \ldots}{1 + \frac{1}{2!}(s\tau) + \frac{1}{4!}(s\tau)^2 + \ldots}$$

$$y_i = \frac{sC_\pi}{g_m} y_f$$

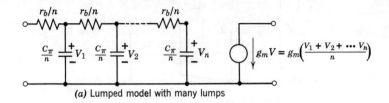

(a) Lumped model with many lumps

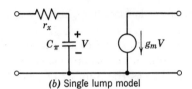

(b) Single lump model

Fig. 8.12.

The approximate model has $y_f$ and $y_i$ of the form

$$y_f = g_m \frac{1}{1 + sr_xC_\pi}$$

$$y_i = sC_\pi \frac{1}{1 + sr_xC_\pi}$$

(a) One common method of choosing $r_x$ is to make the low-frequency phase shift of $y_i$ and $y_f$ for the model agree with these values for the distributed structure. What value of $r_x$ would you use to satisfy this criterion?

(b) An alternate choice is to pick $r_x$ so that the low-frequency magnitude of $y_i$ and $y_f$ matches the true value. What value of $r_x$ would this require?

(c) Find a lumped-pole approximation to $y_f$. If a transistor is accurately modeled by the distributed structure in Fig. 8.12a, what are the approximate bandwidth, $f_h$, and rise time, $T_r$, when the transistor is excited by a voltage source and drives a resistive load?

**P8.2** The $RC$ ladder structure in Fig. 8.13 is a lumped model for describing the high-frequency transport of minority current in the base of a transistor. The following questions have to do with finding approximations for the frequency response of the current gain of this circuit.

(a) For $n = 5$, compute the $r_{ij}$ matrix and the $\tau_{ij}^2$ matrix. Use these to find $T_1$ and $T_2$.

(b) Express $T_1$ and $T_2$ as a function of $n$ and show that for large $n$, $T_1 = RC/3$ and $T_2^2 = (RC)^2/6$.

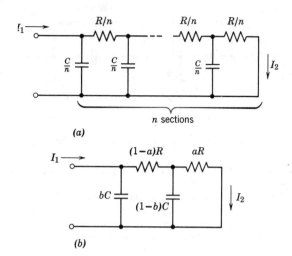

Fig. 8.13.

(c) For the distributed circuit (i.e., large $n$) it can be shown that $I_2/I_1 = (\cosh \sqrt{sRC})^{-1}$. Expand $\cosh \sqrt{sRC}$ in powers of $s$ and compute $T_1$ and $T_2$ from the coefficients of $s$ and $s^2$. Show that the results agree with (b). Also, find the pole locations (there are an infinite number) and show that $T_1 = \Sigma \tau_k$, $T_2{}^2 = \Sigma \tau_k{}^2$. (*Hint:* $\cosh jx = \cos x$.)

(d) Find a lumped-pole approximation to $I_2/I_1$ which is valid for large $n$. Use this lumped-pole approximation to estimate $\omega_h$ and compare with the actual value of $\omega_h = 2.433(RC)^{-1}$.

(e) Find values of $a$ and $b$ in the two-lump model of Fig. 8.13b such that $T_1$ and $T_2$ are the same for the model as for the structure of Fig. 8.13a with $n$ large. Compute $\omega_h$ for this two-lump model and compare with the result of (d).

**P8.3** This problem considers bandwidth calculations for the circuit of Fig. 8.14. Assume $R_s = 50$ ohms, $r_x = 50$ ohms, $r_\pi = 100$ ohms, $g_m = 0.2$ mho, $R_L = 50$ ohms, $C_1 = 10$ pf, $C_2 = C_3 = 1$ pf.

(a) Compute $\tau_{j0}$ and the $\tau_{ij}^2$ matrix.

(b) Show that $T_1 = 1.44$ nsec and $T_2 = 1.38$ nsec, and use these values to estimate the bandwidth of $V_2/V_1$.

(c) Which capacitor has the dominant effect on bandwidth?

(d) Show that the values of $T_1$ and $T_2$ imply a single dominant natural frequency and calculate its location.

(e) Give a physical reason for why $C_3$ has a more pronounced effect on bandwidth than $C_2$.

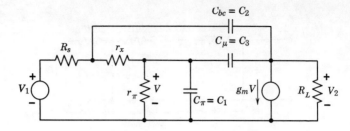

**Fig. 8.14.** High-frequency, single-stage amplifier model.

**P8.4** Assume that there are $n$ equi-spaced poles, as shown in Fig. 8.15, and that all zeros are at infinity.

(a) For $n = 3$ write $H(s)$ and expand in partial fractions. Then evaluate $h_o(t)$ and $h_{-1}(t)$. Assuming $a = b/2 = 1/\tau$, compute $\omega_h$ and $T_r$ and compare with the Gaussian approximation.

(b) Show that for any $n$ the impulse response is

$$h_o(t) = u_1(t) \frac{a(a/b + n - 1)!}{(a/b!)(n-1)!} e^{-at}[1 - e^{-bt}]^{n-1}$$

(c) If $a = b/2 = \tau^{-1}$, then for large $n$, $T_1 = \tau[\ln 2n^{1/2} - \gamma/2]$ where $\gamma = 0.577\ldots$, and $T_2 = \tau \pi^2/8$. What is the significance of the fact that as $n \to \infty$ $T_1 \to \infty$ while $T_2$ remains finite?

(d) For $a = b/2 = \tau^{-1}$ and $n$ large, evaluate $\omega_h$ and compare with (a) and with the value calculated by the Gaussian approximation. (*Hint:* Show that $|H(j\omega)|^2 = [\cosh{(\pi/2)\omega\tau}]^{-1}$).

(e) For $a = b/2 = \tau^{-1}$ and $n$ large, compute $h_{-1}(t - T_1)$ and evaluate $T_r$. Compare with (a) and with the Gaussian approximation.

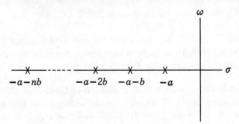

**Fig. 8.15.** Equi-spaced pole configuration.

# Index

Adjustments, balance, 198
  zero, 198
Admittances, input, 89
  output, 89
Advantages of silicon, 171
Aging effects, 200
Alignability, 232
Amplifiers, broadband, 147
  dc, 171
  feedback, 104
  marginally stable, 128
Analysis in feedback form, 81
  of small changes, 191
Asymmetry, 213
Avalanche diode, 178

Balance adjustments, 198
Bandwidth, 8, 151
  for lumped-pole approximation, 276
  of interior stage, 151
  per-stage, 151
  reduction factor for $n$ identical poles, 256
Basic amplifier, 64
Bode plot, 28, 47, 128

Bounds on response, 259
Break away point, 114
Broadband amplifiers, 147
Butterworth filter, 123
Bypass capacitors, 22

Calculating magnitude and phase, from factored polynomial, 47
  without factoring, 49
Calculation of 0.707 point, 17, 45
Capacitors, bypass and coupling, 22
Cascaded stages, 172
Center of gravity, 114, 270
Change, drift, 199
  environmental, 199
  in transistor characteristics, 195
Characteristic times, 246
Circuits, two-transistor, 172
Common-mode positive feedback, 206
Compensation, 118, 120, 138
Complex pole-pair, 106, 114
Composite transistors, 174
Conditions conducive to the feedback viewpoint, 84
Conformality, 128

## Index

Connection, Darlington, 174
Constant-current source, 179
Constant-resistance cascade, 165
Construction of approximate loci, 111
Coupling capacitors, 22
Cramer's rule, 13
$C_t$ approximation, 22, 26
Current-ratio feedback, 78, 96

Darlington connection, 174
dc amplifier, 171
dc feedback, 180
Delay, 250, 257
Desensitivity, 64, 65, 86, 96, 120
Determinant, six-by-six, 28
Differential amplifier, 210
Diode, avalanche, 178
Direct-coupled ac amplifiers, 180
Direct-coupled operational amplifier, 222
Distortion, 69
Dominant natural frequencies, 116
Drift, 200
Droop, 278

Emitter resistance, 158
Environmental changes, 199
Error signal, 63

Factors, mismatch, 234
Feedback, 62
  amplifier configurations, 75
  amplifiers, 104
  current-ratio, 78, 96
  dc, 180, 183
  in linear systems, 71
  negative, 63, 66, 75
  network, modification, 120
  positive, 75
  series, 94
  transadmittance, 75, 91, 115
  transimpedance, 94
  voltage-ratio, 78
Fidelity, 70
Filter, Butterworth, 123

Finding $a_1/a_0$ and $a_{n-1}/a_n$ by inspection, 13
Finding transfer function by nodal analysis, 42
Flow graph, 44
Four-stage feedback amplifier, 134
Frequency, dominant natural, 14, 116
  half power, 45, 246
  highest natural, 20
  lowest natural, 20
  moments, 277
  response, 104, 110
  transverse cutoff, 6, 34

Gain, mid-frequency, 6
Gain and bandwidth performance, 154
Gain margin, 131
Gain versus per-stage bandwidth, 153
Gaussian approximation, 253, 257
Gaussian response, 254
Generator, perturbation, 197

Half-circuit, 203, 209
Half-power frequency, 246
$h_{FE}$, 171
Highest natural frequency, 20
High-frequency poles, 5, 30
High-pass step response, 277
Homologous parameter, 213
Hybrid-$\pi$ model, 5, 34, 39

Impedance effects, 80
Impedance levels, 78
Impulse response, 249, 257, 270
Independent variables, 82
Inductive shunt peaking, 269
Input admittance, 89, 90, 231
Interaction, 224
Interior stage, 147
Internal distortion, 71
Interstage coupling, 176
Intrinsic hybrid-$\pi$ model, 5, 39
Inverse feedback, 63

## Index

Level, impedance, 78
Linear phase response, 254
Loading effects, 86
Loading, shunt, 125
Locus, 106
    root, 108, 110, 111
Locus of 0.707 points, 153
Log magnitude, 126
Loop transmission, 89
Lowest natural frequency, 20, 21
Low-frequency poles, 5
Lumped-pole approximation, 273

Magnitude, log, 126
Marginally stable amplifier, 128
Mid-frequency gain, 6
Mismatch, 236
    to achieve alignability, 236
Mismatch factor, 233, 234
Model, hybrid-$\pi$, 5, 34, 39
    including $R_e'$, 52
Modifications of basic amplifier, 118
Modification of feedback network, 120
Multiple-order poles, 12

Natural frequencies, 14
    dominant, 116
    substitute, 26
Negative feedback, 63, 66, 75
Neutralize, 225
Nichols Chart, 133, 138
Nondominant poles, 124
Nonlinearities, 68
Nyquist criterion, 126

$\omega_h$, 10, 12, 17, 25, 250
$\omega_{45°}$, 250
Open-circuit resistance, 13
Open-circuit time constant, 16, 20
Open-loop gain, 89
Operating point, 182
Operational amplifiers, 101
Output admittance, 89
Over-all bandwidth, 227
Overshoot, 250

$\pi$ model, 33, 34
Parameter changes, 185
Partial fractions, 248
Peaking, 131
Per-stage bandwidth, 151
Perturbation generator, 188, 197
Phase, 50
Phase bounds, 261
Phase margin, 131
Pole locations, 18, 45
    high-frequency, 5, 30
    low-frequency, 5
    multiple-order, 12
    nondominant, 124
Pole-zero diagram, 228
Positive feedback, 75
Power gain, 235
Power supply variations, 217
Precision attenuator, 63

Quiescent operating conditions, 182

Radius of gyration, 270
Reduction of distortion, 69
Relations between $T_2{}^2$ and circuit time constants, 262
Resistance, open-circuit, 13
Resistive broadbanding, 150
Response, frequency, 104, 110
    Gaussian, 254
    impulse, 257, 270
    linear phase, 254
    steady-state, 125
    step, 250
    transient, 110, 130
Rise time, 250, 257, 276
Root locus, 108, 110, 115

Sag, 278
Selectivity, 228
Self-compensation of temperature sensitivity, 219
Series feedback, 94
Series-series, 78
Series-shunt, 78
Short-circuit time constants, 21, 22

Shunt loading, 125
Shunt peaking, 163, 269
Shunt-series, 78
Shunt-shunt, 78
Signal flow graphs, 44
Silicon, advantages of, 171
Six-by-six determinant, 28
Small changes in a nonlinear circuit, 193
Stability, 104
Steady-state response, 125
Step response, 250
 high-pass, 277
Substitute natural frequencies, 26
Sum and difference components, 207
Symmetric circuits, 201
Synchronous tuning, 227

Temperature variations, 219
Thermal and shot noise, 200
Time constants, 16
 open-circuit, 16, 20
 short-circuit, 21, 22
Time domain interpretation of $T_1$ and $T_2$, 270
Time moment, 244, 251, 252, 268

Time response, 245
Transadmittance feedback, 75, 91, 115
Transient response, 110, 130
Transimpedance, 78
Transimpedance feedback, 94
Transistor models, 56
Transistors, composite, 174
Transverse cutoff frequency, 6, 34, 157
True $\pi$ model, 34
Two-transistor circuits, 172

Unbypassed emitter resistor, 52
Undershoot, 250
Unilateral, 225
Upper 0.707 frequency, 10

Virtual ground, 203
Voltage-ratio feedback, 78

$y$-parameter, 35

Zero adjustments, 198
Zeros in $a(s)$, 125